POOR PEOPLE, POOR PLACES

A Geography of Poverty and Deprivation in Ireland

POOR PEOPLE, POOR PLACES

A Geography of Poverty and Deprivation in Ireland

Edited by
Dennis G. Pringle
Jim Walsh
Mark Hennessy

Oak Tree Press
Dublin

*In Association with the
Geographical Society of Ireland*

Oak Tree Press
Merrion Building
Lower Merrion Street
Dublin 2, Ireland
www.oaktreepress.com

© 1999 Individual Contributors and
the Geographical Society of Ireland

A catalogue record of this book is
available from the British Library.

ISBN 1 86076 108 9

All views expressed in this book are the views of the authors, and do not necessarily reflect the views of the editors, the Combat Poverty Agency, the Geographical Society of Ireland or the National University of Ireland, Maynooth.

All rights reserved. No part of this publication may be reproduced or transmitted in any form or by any means, including photocopying and recording, without written permission of the publisher. Such written permission must also be obtained before any part of this publication is stored in a retrieval system of any nature. Requests for permission should be directed to
Oak Tree Press, Merrion Building,
Lower Merrion Street, Dublin 2, Ireland.

Printed in the Republic of Ireland
by Colour Books Ltd.

Acknowledgements

The editors gratefully acknowledge the financial support provided by the Combat Poverty Agency towards the costs of running the conference upon which the contents of this book are based and also towards the costs of publication. They would also like to acknowledge the financial support provided by the National University of Ireland, Maynooth, towards the costs of publication.

They thank the individual contributors, Jim Keenan, the cartographer in the Geography Department in NUI, Maynooth, and David Givens and Jenna Dowds of Oak Tree Press, whose interest and co-operation make this publication a reality.

Finally, they acknowledge the support and assistance of the Geographical Society of Ireland, under whose auspices the annual conference was held (at which most of these papers were presented) and this book is published.

CONTENTS

Acknowledgements .... *v*
List of Contributors .... *xv*

1. **Poor People, Poor Places: An Introduction** 1
    Dennis G. Pringle and Jim Walsh

PART 1. THE SPATIAL DISTRIBUTION OF POVERTY IN IRELAND

2. **Affluence and Deprivation: A Spatial Analysis based on
    the 1991 Census of Population** 13
    Trutz Haase

3. **Spatial Aspects of Poverty
    and Deprivation in Ireland** 37
    Brian Nolan, Christopher T. Whelan and James Williams

4. **Methodological Issues in a Cross-border Investigation
    of Poverty in Ireland** 75
    Sally Cook, Adrian Moore and Michael Poole

PART 2. RURAL POVERTY

5. **Rural Poverty:
    A Political Economy Perspective** 97
    Hilary Tovey

6. **Integration and Exclusion in Rural Ireland** 123
    James A. Walsh

7. **Poverty and Accessibility to Services
    in the Rural West of Ireland** 141
    Mary Cawley

8. **Residents' Perspectives of Rural Living Conditions in Cork and Kerry** 157
   David Storey

PART 3. URBAN POVERTY

9. **Deconstructing Urban Poverty** 177
   Andrew MacLaran

10. **Poor People or Poor Place? Urban Deprivation in Southhill East, Limerick City** 203
    Des McCafferty

11. **Spatial Planning and Poverty in North Clondalkin** 225
    Brendan Bartley

PART 4. POLICY RESPONSES

12. **Something Old, Something New: Lessons to be Learnt from Previous Strategies of Positive Territorial Discrimination** 263
    Dennis G. Pringle

13. **The Role of Area-Based Programmes in Tackling Poverty** 279
    Jim Walsh

PART 5. CONCLUSION

14. **Poor People, Poor Places: Conclusion** 313
    Dennis G. Pringle

LIST OF TABLES AND FIGURES

Tables

2.1	Comparison of the Most Widely-used Deprivation Indices in Britain and Ireland	19
2.2	Structure Matrix of Deprivation Factors Based on Multivariate Analysis of 1991 Small Area Population Statistics	31
3.1	Risk and Incidence of Poverty by Type of Area, 1994	42
3.2	Risk and Incidence of Poverty by Tenure Type, 1994	44
3.3	Risk and Incidence of Poverty in 1994 Using Combined 60% Relative Income plus Basic Deprivation Classified by Area and Tenure Type	45
3.4	Risk and Incidence of Poverty by Planning Region, 1994	47
3.5	Rates and Incidence of Higher and Lower Professional and Unskilled Manual Social Classes by County, 1991	49
3.6	National Distribution of Persons in Higher and Lower Professional and Unskilled Manual Classes Classified by Relevant Decile Rate at District Electoral Division Level in 1991	54
3.7	Distribution within Dublin City and County of Persons in Higher and Lower Professional and Unskilled Manual Classes Classified by Relevant Decile Rate at the Ward/DED Level in 1991	59
3.8	Rates and Incidence of Unemployment by Planning Regions and Counties	60

3.9	National Distribution of Unemployed Classified by Decile Unemployment Rate at District Electoral Division Level in 1991 63
3.10	Distribution of Unemployment Rates in the 322 Wards of Dublin City and County in 1991 65
3.11	Distribution of the Unemployed in Dublin City and County Classified by Decile Unemployment Rate at the Ward/DED Level in 1991. 67
4.1a	Data Limitations Affecting Indicator Variables 79
4.1b	Scale Issues in Area Analysis 80
4.2	Census-derived Variables Used as Deprivation Indicators in Great Britain and/or Ireland 80
4.3	Indicators of Housing Deprivation — Percentage of Households Affected in Northern Ireland, the Republic of Ireland and the Border Regions 87
8.1	Non-possession of Selected Household Amenities in Study Areas 160
8.2	Percentage of Family Units in Each Category Claiming to be Poor 164
8.3	Percentage in Each Socio-economic Group Claiming to be Poor 165
8.4	Causes of Poverty Cited by Respondents 168
8.5	Respondents' Perceptions of Causes of Poverty 171
9.1	Shrinkage of Industrial Land Uses, 1966–85 191
10.1	Population Distribution by Age Group 208
10.2	Stage in Family Cycle 209

10.3	Lone Parent Young Families (all children under 15 years)	210
10.4	Age- and Sex-Specific Rates of Labour Force Participation	211
10.5	Age-Specific Unemployment Rates	212
10.6	Annual Percentage Rates of Population Change, 1971-1991	217
10.7	Age-and Sex-specific Levels of Differential Migration, 1981-91	219
11.1	Age Structure — Ireland, Dublin North, Clondalkin	237
11.2	North Clondalkin Neighbourhoods and Estates	238
11.3	Profile of Areas of Need	239
11.4	Differing Explanations of Urban Problems (after Bennington, 1979)	248
12.1	Percentage of Different Types of Social Problem which would be Enclosed in Designated Areas of Different Sizes in Belfast	271

Figures

2.1 Affluence and Deprivation in Ireland, 1991 46

3.1 Per Cent Higher and Lower Professional, 1991 51
3.2 Per Cent of Unskilled Manual, 1991 52
3.3 Dublin County and County Borough; Percentage in Higher and Lower Professional, 1991 56
3.4 Dublin County and County Borough; Percentage in Unskilled Manual Class, 1991 57
3.5 Unemployment Rates, 1991 61
3.6 Dublin County & County Borough: Unemployment Rates, 1991 66

4.1 Overlap of Deprivation Indicators for Northern Ireland and the Irish Republic 85
4.2 Percentage of Households in Public Ownership 87
4.3 Percentage of Households Lacking a Bath or Shower 88
4.4 Impact of Urban Scale Differences in Cross-border Comparative Analysis: Male Unemployment Levels for (a) Coleraine, NI and (b) Dundalk, RoI 90

8.1 Map of Cork and Kerry Showing Study Areas 159
8.2 Number of Amenities Lacked 161
8.3 Self-defined Poor Classified by Number of Amenities Lacked 163
8.4 Poverty Perceived as Problem Classified by Amenities Lacked 167

9.1	Shrinkage of Industrial Land Uses, 1966–85	189
10.1	Ward Deprivation Scores, 1991, Limerick County Borough	205
10.2	Location Map	206
10.3	Differential Migration, 1981–91	218
11.1	Population Growth, 1961–89	232
11.2a	Age Structure — Dublin	233
11.2b	Social Class Structure — Dublin	234
14.1	The Poverty/Deprivation Nexus	319

LIST OF CONTRIBUTORS

Brendan Bartley is a chartered town planner and chartered transportation planner, with experience in the public and private sectors. He currently lectures in urban geography and town planning in NUI, Maynooth, where he directs the Land Use and Transportation Unit, a research centre specialising in socio-economic, land-use and transportation studies. He is currently working on a research project on urban social exclusion, funded by the Combat Poverty Agency and the Katharine Howard Foundation. At an EU level, Brendan is involved in two transportation planning research networks: Mobility Impact Reaction and Opinions (MIRO) and Vade Mecum; and two targeted social and economic research projects on urban planning, Urban Redevelopment and Social Polarisation in the City (URSPIC), and Social Exclusion in European Neighbourhoods. He has recently published in *The Changing Social Environment* (Oscail/Dublin City University) and in Madanipour *et al.*, (eds.) *Social Exclusion and the Neighbourhood in Europe* (Jessica Kingsley).
E-mail: bbartley@may.ie

Mary Cawley is a senior lecturer in geography at the National University of Ireland, Galway. Her research interests and publications relate to aspects of the social and economic geography of rural areas. Access to services forms one focus of her research. She recently completed a three-year term (1996–99) as president of the Geographical Society of Ireland.
E-mail: mary.cawley@nuigalway.ie

Sally Cook is a lecturer in geography at the University of Ulster. Her research interests include: medical geography (health care provision and uptake, spatial epi-demiology, and impact of deprivation on health); social geography (spatial patterns of deprivation and methodology of multi-state comparison); geographical methodology (impact of scale and settlement size on area-based spatial analysis); and applications of GIS. Sally's recent publications are *Deprivation in the Irish Border Region: a Census-Based Study*, a report for the Northern Ireland Voluntary Trust, with M.A. Poole, D.G. Pringle and A.J. Moore (University of Ulster, 1997); "Attitudes to the Countryside" in G. Robinson *et al.*, (eds.) (1998), *Social Attitudes in Northern Ireland, The Seventh Report*, with A.J. Moore and C.F. Guyer, (Appletree Press, Belfast); "Attitudes to the Environment in Northern Ireland" in G. Robinson *et al.*, (eds.), (1998) *op cit*, with A.J. Moore and C.F. Guyer; and *Comparative Spatial Deprivation in Ireland*, with M.A. Poole, D.G. Pringle and A.J. Moore, a report for the Combat Poverty Agency (University of Ulster, 1998).
E-mail: s.cook@ulst.ac.uk

Trutz Haase is an independent social and economic consultant. He formerly worked as a research officer in the Northern Ireland Economic Research Centre and in the Combat Poverty Agency, and as a research associate in the Educational Research Centre at St Patrick's College, Drumcondra. Trutz has undertaken extensive evaluations of local development programmes for Area Development Management, the Irish Government and the EU. He is best known for his work on the development of an Irish index of relative affluence and deprivation and was joint editor of *Poverty in Rural Ireland* (Oak Tree Press/Combat Poverty Agency, 1996). Recent work includes a community uptake analysis of the EU Special Support Programme for Peace and Reconciliation in Northern Ireland.
E-mail: thaase@iol.ie

Mark Hennessy is a lecturer in geography at Trinity College Dublin and a committee member of the Geographical Society of Ireland.
E-mail: mhnnessy@tcd.ie

List of Contributors

Andrew MacLaran is head of the Department of Geography and joint director of the Centre for Urban and Regional Studies at Trinity College Dublin. His background is in welfare geography, including doctoral research on spatial variations in living conditions in Dundee. His research interests lie in the urban arena, particularly the social consequences of economic restructuring and the impacts of the property development sector on the urban environment.
E-mail: amclaran@tcd.ie

Des McCafferty is head of the Department of Geography at Mary Immaculate College, University of Limerick. His main research interests lie in regional and local development, particularly policy and practice in relation to area-based initiatives to combat social exclusion. Recent publications include the book *Local Partnerships for Social Inclusion?* with Jim Walsh and Sarah Craig (Oak Tree Press/Combat Poverty Agency, 1998), and the edited volume *Competitiveness, Innovation and Regional Development in Ireland* with James A. Walsh. The latter is published by the Irish Branch of the Regional Studies Association, of which he is currently chairman.
E-mail: des.mccafferty@mic.ul.ie

Adrian J. Moore is a lecturer in geography at the University of Ulster. His research interests include geographic information systems, medical geography, spatial epidemiology, deprivation and health, accessibility and utilisation of health care services and geodemographics. His recent publications are "Deprivation Payments in General Practice: Some Spatial Issues in Resource Allocation in the United Kingdom", *Health and Place*, 1(2), 1995; "Attitudes to the Countryside", in G. Robinson *et al.*, (eds.), *Social Attitudes in Northern Ireland, The Seventh Report*, (Appletree Press, Belfast, 1998); "Attitudes to the Environment in Northern Ireland", with S. Cook and C.F. Guyer, in Robinson G. *et al.*, (eds.), *op cit*; and *Comparative Spatial Deprivation in Ireland*, with S. Cook, M.A. Poole, D.G. Pringle, a report for the Combat Poverty Agency (University of Ulster, 1998).
E-mail: a.moore@ulst.ac.uk

Brian Nolan is a research professor at The Economic and Social Research Institute. He has published extensively in the areas of poverty, income inequality, tax and social welfare policy, and the labour market. He is joint editor of *Poverty and Policy in Ireland* (Gill and Macmillan, 1994), and co-author of *Resources, Deprivation and Poverty* (Clarendon Press, 1996) and *Poverty in the 1990s* (Oak Tree Press/Combat Poverty Agency, 1996).
E-mail: brian.nolan@esri.ie

Michael A. Poole is a senior lecturer in Environmental Studies, University of Ulster at Coleraine. His research interests are: methodological issues in spatial analysis, especially the scale problem; residential segregation analysis, both empirical and methodological; political violence and ethnicity in Northern Ireland; and small area analysis of Irish deprivation data. Michael's Recent publications are "The Spatial Distribution of Political Violence in Northern Ireland: An Update to 1993", in A. O'Day (ed.), *Terrorism's Laboratory: The Case of Northern Ireland* (Dartmouth Publishing Company, 1995), *Ethnic Residential Segregation in Belfast*, with P. Doherty (Centre for the Study of Conflict, University of Ulster at Coleraine, 1995); *Ethnic Residential Segregation in Northern Ireland*, with P. Doherty (Centre for the Study of Conflict, University of Ulster at Coleraine, 1996); "In Search of Ethnicity in Ireland", in B. Graham (ed.), *In Search of Ireland: A Cultural Geography* (Routledge, 1997); "Political Violence: The Overspill from Northern Ireland", in A. O'Day (ed.), *Political Violence in Northern Ireland: Conflict and Conflict Resolution* (Praeger, 1997); and "Ethnic Residential Segregation in Belfast, Northern Ireland, 1971–1991", with P. Doherty, *Geographical Review*, 87(4) (1997).
E-mail: ma.pool@ulst.ac.uk

Dennis G. Pringle is a senior lecturer in geography at the National University of Ireland, Maynooth. His main research interests are in medical geography, especially social and spatial inequalities in health; political geography; and computer applications in geography (such as Geographical Information Systems). Dennis is a former president, secretary and membership secretary of the Geographical Society of Ireland. Recent publications include

articles in the international journals *Social Science and Medicine* and *Political Geography*, and (with Sally Cook, Adrian Moore and Mike Poole) reports on cross-border studies of poverty and deprivation submitted to the Northern Ireland Voluntary Trust and the Combat Poverty Agency. He is currently editing special thematic issues on nationalism and national conflicts of the journals *Geopolitics*, *Geography Research Forum* and *GeoJournal*.
E-mail: dpringle@may.ie

David Storey is a senior lecturer in the Department of Geography, University College Worcester. His research interests are rural living conditions, access and mobility in rural areas, rural development, geopolitics, nationalism, and images of place. David has published articles on service provision, social change and rural development in Ireland. His forthcoming book is *Territories: Ways of Dividing the World*, (Addison Wesley Longman). Also forthcoming is an article entitled "Issues of Integration, Participation and Empowerment in Rural Development: the Case of LEADER in the Republic of Ireland", *Journal of Rural Studies* 15(3).
E-mail: d.storey@worc.ac.uk

Hilary Tovey is a senior lecturer in sociology at Trinity College Dublin, and an executive committee member of the European Society for Rural Sociology. Her research interests include food production and consumption, alternative agricultures, rural society and environment, and issues in minority cultures, particularly language. Recent publications include papers on organic farming in Ireland in *Sociologia Ruralis* (1997) and the *Irish Journal of Sociology* (1999), a chapter in the book *New Actors on the European Countryside* (Budapest, 1998), and co-editorship of the Combat Poverty/Oak Tree Press publication *Poverty in Rural Ireland* (1996).
E-mail: htovey@tcd.ie

James A. Walsh is professor and head of the Geography Department at the National University of Ireland at Maynooth. He has written extensively on regional and local development in Ireland. Recent publications include *Competitiveness, Innovation and Regional Development in Ireland*, co-edited with Des McCafferty (Regional Studies Association (1997)) and *Sustainable Development on the North Atlantic Margin*, co-edited with R. Byron and P. Breathnach (Ashgate (1997)). He was a member of the Rural Development Policy Advisory Committee 1995–97 and is currently a member of the National Economic and Social Council. He has undertaken research on local and regional development for the Irish Government, the EU and the OECD.
E-mail: jawalsh@may.ie

Jim Walsh is a researcher with the Combat Poverty Agency, where his duties have included managing the Agency's research programme and preparing policy submissions. His research interests include the geography of poverty, local and community development initiatives, and access to financial services/indebtedness. He is the principal author of *Local Partnerships for Social Inclusion?*, the Irish element of a cross-national study by the European Foundation for the Improvement in Living and Working Conditions (Oak Tree Press/Combat Poverty Agency, 1998). Previously, Jim was the research and evaluation officer with the PAUL Partnership in Limerick (1991–94), where he authored the final evaluation report on the partnership's involvement in the Poverty 3 programme (PAUL Partnership, 1994).
E-mail: walshj@cpa.ie

Christopher T. Whelan is a research professor at The Economic and Social Research Institute. He is joint author of *Understanding Contemporary Ireland* (Macmillan, 1990) *Social Mobility and Social Class in Ireland* (Gill and Macmillan, 1994); and *Resources, Deprivation and Poverty* (Clarendon, 1996). He is joint editor of *Values and Social Change in Ireland* (Gill and Macmillan, 1994), *The Development of Industrial Society in Ireland* and *Ireland North and South: Perspectives from Social Science* (OUP for the British Academy, 1992 and 1999).
E-mail: chris.whelan@esri.ie

James Williams is a senior research officer and head of the survey unit at The Economic and Social Research Institute. His recent publications have been on income distribution, poverty and survey techniques. His main research interests include survey methodology and sample design, poverty and income distribution, geographic information systems and regional development.
E-mail: james.williams@esri.ie

1

POOR PEOPLE, POOR PLACES:
AN INTRODUCTION

Dennis G. Pringle
National University of Ireland, Maynooth

Jim Walsh
Combat Poverty Agency, Dublin

The British Conservative government under Harold Macmillan in the early 1960s was fond of telling the British electorate that "You have never had it so good". The same could be said of Ireland in the mid-1990s. Ireland is currently experiencing an unprecedented wave of prosperity, popularly referred to as the "Celtic Tiger". However, not everyone living in Ireland has shared equally in this prosperity. As one Maynooth student poignantly complained in response to a recent exam question on the Celtic Tiger: "West of the Shannon, people have not even seen as much as a fluff-ball". The same could be said of large tracts of public sector housing in Dublin and the other major cities, whilst there are doubtless many other areas throughout the state which have been bypassed to the same extent by the recent prosperity. Even within those areas which have benefited from the economic boom, it is obvious that some people are benefiting much more than others. Recent research by the ESRI suggests that one-in-five households fall below an income poverty line of 50 per cent of average income (Callan *et al.*, 1996).

Social inequalities have always been with us, and presumably always will be. It may therefore appear anomalous to be expressing concern about poverty at a time when the national economy is so obviously booming. However, if certain sections of the population have been bypassed by the recent prosperity, to the extent that they can be regarded as socially excluded, then there are clearly problems of social inequality which need to be addressed. There is no better time for tackling these problems than the present, when we have the resources to possibly do something about the problem. The alternative is to wait until the next economic slump, when not only will the distress of those who are worst off become objectively worse but there will be fewer resources available for remedial action.

Poverty and social inequality are, by definition, social problems, but the processes that generate these inequalities do not take place in a vacuum, nor do they operate in a uniform manner throughout the entire state. Rather, they tend to produce different outcomes in different areas in response to local conditions. These conditions vary not only with regard to the inherent characteristics of each area (e.g. natural resources, location relative to other places), but also with regard to various other attributes determined by the cumulative historical legacy of past social, economic, cultural, political and administrative processes (e.g. social composition, landholding patterns, accumulated wealth and capital availability, physical infrastructure). Each place, in short, is unique, with the result that processes originating at a global, European or national level will tend to have a different, or at least differential, impact in different areas. The processes generating social inequalities may be primarily "social" and "economic", but they cannot be fully understood unless they are located within a historical and geographical context.

Geographers consequently have an important contribution to make to the understanding of poverty and social inequality. Poverty is obviously not uniformly distributed throughout the state — i.e. some areas clearly experience much higher levels of poverty than other areas. However, the spatial dimensions of poverty in Ireland are still only poorly understood. For example, are there marked regional disparities in the percentages of people experiencing poverty? Is poverty more common in urban areas or rural areas? Is the experience of poverty affected by the environment in which people live, especially in areas that either have a concentration of poverty or may be categorised as peripheral and remote? Do the processes which cause social inequalities operate differently in urban and rural areas? Do different types of process operate at different spatial scales? To what extent do apparently affluent areas disguise hidden pockets of poverty? Can space influence the transmission of poverty across generations?

The problem is not that geographers do not have answers to these questions — they do. The problem is that the answers provided by geographers are not necessarily consistent. Few geographers in Ireland have regarded "poverty" as the central object of their research; rather poverty for most geographers has tended to emerge as a "side issue" in research primarily focused on other topics (e.g. processes of urban change, processes of rural change, spatial inequalities in health, local and regional development). The answers provided by geographers to these questions consequently tend to be coloured by their other research interests. This book represents an attempt to pull some of these disparate sources of information on the geography of poverty together in a single volume and in the process hopefully initiate a debate from which a consensus may eventually emerge.

The need for a consensus on the spatial dimensions of poverty is given additional weight by trends in recent years towards an increasing use of local area-based strategies as a policy instrument for delivering services and resources. Area-based strategies offer a number of potential advantages over more traditional centralised administrative systems: they provide a mechanism for allocating additional resources to areas of special need; they facilitate possible closer co-ordination at a local level of services provided by different government departments at national level; they permit greater flexibility with regard to tailoring policies to specific local needs; and they provide the potential for the views of local people to be taken into account through partnership. Nevertheless, the introduction of area-based strategies, whilst to be welcomed, raises a number of issues which require further scrutiny. For example, how should areas of special need be identified? How should the problems in these areas be tackled? How does one avoid discriminating in favour of affluent people in areas of need at the expense of poor people in areas which are not designated? How does one ensure that the wishes of local people are taken into account? How much power should be delegated from central to local level? What should be the relationship between central government and local areas? Are different strategies required for urban and rural areas? Similar questions arise in the context of the proposed division of Ireland into two regions for the structural funds. These are all issues on which geographers should be able to make a contribution. Yet, ironically, discussions about the design and implementation of area-based approaches have largely excluded the professional practitioners of geography. This clearly represented a missed opportunity to shape policy for a discipline with as strongly an applied focus as geography.

It is against this background that the Geographical Society of Ireland (GSI) decided to hold a conference entitled "Poor People, Poor Places" in September 1996 in the National University of Ireland, Maynooth. The Geographical Society of Ireland has an established tradition of organising a one-day conference each year at a different venue on a selected theme of topical interest. Previous conferences have focused on topics such as planning in Dublin, the Irish Sea as a resource base, rural development, migration, and the impact of the European Union. These conference are targeted at a broad-based audience, especially policy makers employed in the public service, with speakers normally selected from active researchers within the geographical community. Poverty was selected as the theme for the 1996 conference, which was organised on behalf of the GSI by the editors of the present volume (i.e. Dennis Pringle, National University of Ireland, Maynooth; Jim Walsh, Combat Poverty Agency; and Mark Hennessy, Trinity College, Dublin). The conference attracted an audience of over 100 people from all parts of the country, and it included representatives from government departments, state agencies, local government politicians, partnerships and community groups, along with teachers, academics and other concerned individuals. The Combat Poverty Agency provided financial support, which enabled a number of bursaries to be provided.

Given the very positive feedback from those attending the conference, it was decided to explore the possibility of publishing the proceedings in order to make the material available to a wider audience. This took much longer than had originally been anticipated, but this book represents the fruits of our labour. Ten of the 14 chapters in this book are based on papers originally presented at the Maynooth conference. Two additional chapters were commissioned from Andrew MacLaran and Hilary Tovey (both Trinity College, Dublin), whilst this short introduction and a more substantial concluding chapter were added by two of the editors (viz. Dennis Pringle and Jim Walsh). In addition, authors were given the chance to revise and update their original papers.

The structure of the book roughly corresponds to the organisation of the paper sessions at the conference. The chapters in the book are organised into four parts (not counting the introductory and concluding chapters).

There is a considerable diversity in the material presented in the book, with empirical analyses, policy reviews, methodological papers, detailed case studies and macro overviews. There is also a diversity in the understanding of the spatial aspects of poverty, with some authors seeing it as having only a residual effect while others place the spatial aspect at the core of their analysis. This is in part related to the eclectic nature of geography, as well as to the fact that some contributors are sociologists, economists or planners. There remains, however, a fundamental coherence to the collection in that all are concerned with the spatial aspects of poverty.

Part One contains three chapters summarising the findings from empirical studies of the spatial distribution of poverty in Ireland. All three studies raise a number of fundamental methodological issues. Trutz Haase (an independent consultant) sets the scene in Chapter 2. Haase outlines a number of conceptual and methodological issues associated with the construction of a multivariate deprivation index. He then reviews the main features of various deprivation indices developed elsewhere before defining his own deprivation index based on 13 indicators proven to be associated with deprivation, including unemployment, underemployment, age dependency, lone parents, social class, education levels and population density. The chapter concludes with a discussion of the spatial patterns identified using this index in the Republic of Ireland.

In Chapter 3, Brian Nolan, Chris Whelan and James Williams (Economic and Social Research Institute) take a quite different approach to the spatial distribution of poor households, using data from an income and lifestyle survey of a representative sample of 4,000 households. The authors use income and resource-based concepts of poverty to examine the location of poor households at a variety of spatial levels (e.g. region, population size, tenure). These findings are then compared with data on the spatial distribution of selected indicators of deprivation from the Small Area Population Statistics derived from the 1991 census returns. This chapter highlights, amongst other things, the importance of housing tenure as a correlate of poverty.

Sally Cook, Adrian Moore and Mike Poole (University of Ulster at Coleraine) discuss some of the methodological problems arising out of a comparative cross-border study of deprivation in Chapter 4, again with a focus on census data. This chapter examines some of the problems associated with measuring deprivation in one state before looking at some of the special problems which arise when attempting to compare the situation in two states. These problems arise because of differences in the nature of the data collected by the northern and southern government agencies; differences in the areal units used by these agencies; and differences in the social, economic, cultural, political and administrative contexts between the two jurisdictions.

Part Two contains four chapters which focus on poverty in rural areas. Hilary Tovey (Trinity College, Dublin) sets the scene for the rural papers in Chapter 5. She begins by raising questions about what is meant by poverty and, more especially, what is meant by "rural". She argues that poverty is a relative condition generated by social inequalities. She then reviews some of the major processes contributing to structural change and the production of social inequalities and poverty in rural areas, including changes in food production/agriculture, the impact of industrialisation and urbanisation, and new forms of rural resource use. These processes are the local manifestations of global capitalist processes. Policies aimed at tackling poverty should attempt to address the social inequalities which these processes create. Tovey provides some suggestions, whilst questioning both the usefulness of local-based strategies and also the uncritical promotion of sustainable development strategies.

James A. Walsh (National University of Ireland, Maynooth) also reviews some of the processes contributing to changing conditions in rural areas such as Europeanisation and globalisation; social and demographic changes; the changing role of the state; and technological changes in Chapter 6. However, the main purpose of his chapter is to look at the current institutional arrangements for rural development, and to identify some of the major weaknesses, before making a number of recommendations for the development of an improved administrative system for rural development and social inclusion based upon partnership and closer vertical and horizontal integration of the relevant institutions.

Mary Cawley (National University of Ireland, Galway) reviews some of the problems caused by poor accessibility encountered by rural dwellers in Chapter 7. Although these problems are by no means confined to poor people, certain sub-groups within the population, including the elderly, those on low income, and mothers without day-time access to private transport are especially disadvantaged. Cawley considers how factors such as rural population decline and policies of economic rationalisation have combined to reduce the provision of essential services in rural areas. Special attention is given to the implications of reduced accessibility to health care, primary and secondary education, and transport provision. The chapter concludes with a brief review of some remedial strategies.

Chapter 8, by Peter Storey (University College Worcester), examines the experience of poverty in a specific rural location, an undercurrent theme in a number of the other chapters. Do people experiencing poverty experience it in different ways? Indeed, do some people experiencing poverty even perceive themselves to be disadvantaged? These are some of the questions addressed in his analysis of the findings from a survey of 191 households in Cork and Kerry. Respondents were asked to identify if they possessed or lacked a number of household items. This provided the basis of an "objective" measure of deprivation. Respondents were also asked if they perceived themselves to be poor, if they believed there was poverty in the local area or elsewhere in Ireland; and what they considered to be the causes of poverty. Storey found a reasonable degree of correspondence between the objective indicators of poverty and people's perception of their own situation, but he also found some interesting contrasts with regard to what people perceived to be the causes of poverty.

Part Three contains three chapters on poverty in urban areas. In Chapter 9, Andrew MacLaran (Trinity College, Dublin) argues that poverty is a relative concept intimately related to social inequality. After briefly reviewing some of the explanations that have traditionally been put forward to explain poverty, MacLaran argues that social inequalities, and hence poverty, are an unavoidable aspect of the capitalist mode of production. He critically reviews the role of the state in the creation and reproduction of social inequalities, as manifested in education, spatial planning and economic development, and questions whether state-initiated area-based policies have the potential to address the underlying causes of poverty or are simply an ideological diversion.

Chapter 10, by Des McCafferty (Mary Immaculate College, Limerick), shifts the focus from the macro to the micro in his examination of deprivation in O'Malley Park and Keyes Park, two of the four estates which make up the Southill area in the southern part of Limerick city. McCafferty examines a number of sociodemographic indicators which highlight the fact that the area is inhabited by a generally youthful population characterised by a high percentage of single parent families, low labour force participation rates and high unemployment rates. Taken together, these indicators suggest the likelihood of a high level of poverty. In the second part of the chapter, McCafferty notes that one of the reasons why the area is characterised by such a youthful and potentially problematic population is that a large percentage of the older (and probably more affluent) population has moved out. Relatively few of the remaining residents have taken advantage of tenant purchase schemes. McCafferty argues that this instability arises because Southill is not perceived as an attractive place to live, due to poor housing quality and the physical layout of the estates — i.e. it is a "poor place". The chapter concludes with a number of suggestions for alleviating some of these problems.

Brendan Bartley (National University of Ireland, Maynooth) develops similar themes in a local study of North Clondalkin in Chapter 11. Bartley reviews the origins of the neighbourhood and transportation planning ideas which influenced the development of the three new towns in west County Dublin. He then looks at some of the problems which have emerged in the Neilstown, Rowleigh and Quarryvale neighbourhoods in North Clondalkin. As in Southill, problems of social deprivation are compounded by a physical infrastructure which is inappropriate to the needs of the residents of these areas. Bartley further contends that the physical design of the North Clondalkin also serves to hide the deprived nature of the area from public gaze. Planners tend to think of themselves as the guardians of the public interest. However, Bartley argues, from a structuralist perspective, in the final sections of the chapter, that planning as a state activity serves to manage societal conflicts over the use of space and thereby helps maintain the *status quo*.

Part Four contains two chapters on policy-related issues. Dennis Pringle (National University of Ireland, Maynooth) reviews some of the experiences of area-based policy strategies in Britain and the US in the 1960s and 1970s in Chapter 12. It is argued that there are important lessons to be learnt from these schemes which are very relevant to some of the strategies being adopted in Ireland in the 1990s. The chapter critically reviews the advantages of area-based approaches with regard to: (a) efficiency in targeting those in need; (b) administrative flexibility, including the potential for community participation in the decision-making process; and (c) effectiveness in counteracting the causes of poverty and social inequality.

Jim Walsh (Combat Poverty Agency) complements Pringle's historical analysis by examining the contemporary upsurge in spatial interventions in Ireland in Chapter 13. Walsh traces the popularity of area-based programmes to a new localism in welfare policy, reflecting a new geography of poverty, a desire for better co-ordination of services, an enhanced role for local communities and support for local employment initiatives. Walsh posits a threefold classification of spatial intervention strategies, which show both similarities and differences to those advanced by Pringle: targeting of resources; enhancement of service provision; and integrated local development. He argues that despite the growing importance of the spatial dimension in anti-poverty strategies, spatial factors still remain poorly understood. He calls for a greater input from geographers in conceptualising the problems of "poor places" and for a stronger spatial dimension in government policymaking.

Part Five contains a single concluding chapter written by two of the editors (Dennis Pringle and Jim Walsh). This represents an attempt to identify some of the major areas of agreement and disagreement highlighted by the other chapters in this book.

The editors would like to acknowledge the support provided throughout the organisation of the conference and the compilation of theses proceedings by the Committee of the Geographical Society of Ireland, especially its President, Dr. Mary Cawley (who is also a contributor to the present volume). We would also like to acknowledge the role played by Celine McHugh, who administered the bookings for the conference, and her assistants who ensured the smooth running of the conference on the day. We would like to thank the Combat Poverty Agency, not only for its financial and administrative support in the organisation of the conference, but also for its financial support for this publication. Finally, we would like to thank the publishers, Oak Tree Press, and each of the authors for the patience which they have shown in bringing this book to fruition. We hope that our readers will agree that the effort has been well worthwhile.

REFERENCES

Callan, T., Nolan, B., Whelan, C. and Williams, J. (1996) *Poverty In the 1990s*, Oak Tree Press, Dublin.

Part 1

THE SPATIAL DISTRIBUTION OF POVERTY
IN IRELAND

2

AFFLUENCE AND DEPRIVATION: A SPATIAL ANALYSIS BASED ON THE 1991 CENSUS OF POPULATION

Trutz Haase
Social and Economic Consultant

BACKGROUND

Poverty, social exclusion and deprivation have received a good share of attention in public discussion in Ireland. However, these concepts are less than clear, often being used interchangeably, and are frequently applied to the general development of an area rather than to the differing situations of the people living within it. The task of defining carefully what exactly we are talking about when we use these terms is therefore of primary importance and precedes both the identification of deprived groups and the development of policies to ameliorate their situation.

Any approach to understanding these issues should reject the assumption that certain areas or occupational groups are poor *per se*, and attempt to understand society in terms of its social class composition and the enormous differences in access to property, capital and means of production that this entails. We also need to take into account the differences which exist in relation to the status that individuals are able to obtain within their communities on this basis. With this in mind, we will begin by discussing a series of contrasts which are frequently made in studies of poverty and social exclusion: (a) absolute versus relative poverty, (b) objective versus subjective approaches and (c) material versus social definitions of deprivation/exclusion.

An *absolute* view of poverty assumes that it is possible to determine in some "scientific" or "value-free" way a minimally acceptable standard of living, often referred to as "basic needs". Few studies of poverty in the developed world would now restrict their definition of poverty to the minimum subsistence levels needed to sustain physical existence. Instead, it is generally agreed that the notion of poverty assumes a distinct meaning in different societies and hence we now refer to it as a relative rather than absolute condition. In Ireland, the latter view is well represented by a number of studies undertaken by the ESRI on the basis of the 1987 *Survey of Income, Poverty and Usage of State Services.*

The distinction between *objective* and *subjective* conceptions of poverty brings to light another dimension: should people's perceptions of their own situation be included in measurements of poverty, or should our analysis be restricted to "external" or "factual" criteria? An emphasis on the subjective dimension may allow us to take into account how people evaluate their own situation rather than assuming that we can understand this perfectly from the outside. Indeed, movement towards more subjectivist approaches have been accompanied by a gradual shift in terminology from "poverty" to "social exclusion", reflecting a concern about how the condition is experienced by those affected. However, it is not clear how subjective criteria could be utilised in the measurement or detection of poverty, particularly as the existing data sources contain objective measures only. In the absence of clear theoretical arguments as to why "subjective" responses should diverge radically from "objective" poverty, it seems acceptable to proceed on an objective basis, using qualitative research as a guide to how our research should be carried out and interpreted.

Whilst economic deprivation and other *material* considerations remain central to defining poverty, it is increasingly believed that policy responses should be formulated in a more holistic manner than this, encompassing not only the material conditions but also the *social* situation in which a person finds him/herself. Whilst the emphasis on the social significance of poverty has allowed this debate to link up with discussions about social justice and the "rights of citizenship", there remains considerable debate about whether it is possible to develop objective measures of social or political *exclusion*, and indeed about whether this can be measured at all. For example, the work of the ESRI has concentrated on refining the *measurement of poverty* and on studying the dynamic processes which may lead people into a situation of poverty, but generally maintains that the framework of *poverty* studies is preferable to that provided by the notion of *social exclusion*.

We would concur with this view whilst insisting on the possibility of a social definition of poverty which avoids the subjectivist pitfalls of the notion of "social exclusion" as well as the false polarity between "material" and "social" definitions. Poverty is an irreducibly social phenomenon which is both objective (i.e. it is an objectively-existing situation with definite causes and therefore measurable either directly or by proxy) and relative (i.e. what constitutes poverty varies across societies and over time). We will thus use the term "social exclusion" to refer to situations where the possibility of active involvement in the social life of a community is restricted for some of its members. This concept does not replace the concept of "poverty" but instead complements it.

The paper is divided into five sections: Sections 1 and 2 outline the relevant methodological considerations which arise during the construction of a census-based deprivation index. Section 3 discusses potential indicators of poverty. Section 4 describes the final selection of indicators used in the Irish Deprivation Index and the final section presents the resulting spatial distribution of affluence and deprivation.

METHODOLOGICAL CONSIDERATIONS

Irish studies concerned with the distribution of income are, by necessity, based on samples of the population. This makes them limited in their usefulness for developing a comprehensive picture of the geographical distribution of poverty and, indeed, there is no available data source whatsoever that would provide information on household incomes at any significant level of spatial disaggregation. For this reason, successive studies of the geographical aspects of income distribution have depended on the use of proxies that can meaningfully be employed to indicate the relative likelihood of poverty in a particular area. The methodological approach of the analysis presented here is based on a multivariate analysis of poverty surrogates from the 1991 Census of Population and builds on previous work by Williams (1993), and Haase (1993, 1995).

The basic consideration informing the work of these authors is that, although unemployment is one of the most significant factors associated with disadvantage, unemployment rates alone do not provide a sufficient indicator of the underlying disadvantage in an area. Indeed, as will become apparent below, high unemployment rates are a predominantly urban phenomenon which, taken on their own, would introduce considerable bias into any nation-wide study. The reason for this is the high level of emigration from many rural areas, which results in the reduction of prevailing unemployment rates. However, as emigration tends to be selective, in that it is concentrated among the working-age population, it tends to leave behind a disproportionately large economically dependent population. Therefore, sustained levels of outmigration can be identified by the age dependency ratio, measured as the proportion of those under fifteen years of age and over sixty-five as a proportion of the total population.

The second reason why long-term adverse labour market conditions may assert themselves in phenomena other than unemployment lies in the peculiarities of small-scale farming. Strong social incentives encourage farmers, even where they do not derive a sufficient family income from farming, to hold on to economically unviable holdings. In such situations, the unemployment rate is likely to understate the real extent of labour market disadvantage, as this is partly concealed in the form of on-farm underemployment.

A number of considerations are important when constructing an index which aims to represent multiple dimensions of deprivation. These are related to the conceptualisation of disadvantage, the choice of variables, the method of analysis, and the combination of indicators to derive a single overall measure. It is also important to consider the purpose for which the index is being constructed, as there may not be a single indicator which is equally appropriate in all circumstances. The following sections briefly outline the thoughts which informed the construction of the present index.

Conceptualising Deprivation

For the purposes of developing a general index, the nature of deprivation must be conceptualised in broad terms, so as to include not only income poverty and its consequences, but also social and environmental problems. These latter issues were seen as particularly important in the context of the development of area-based policies. As Coombes *et al.* (1993) state:

> The fundamental implication of the term deprivation is of an absence . . . of essential or desirable attributes, possessions and opportunities which are considered no more than the minimum by that society (p. 5).

We have already drawn attention to the strong urban bias inherent in the use of the unemployment rate as a single measure of deprivation. This problem is of particular concern in Ireland, as the country as a whole, and rural areas in particular, have experienced long periods of substantial emigration. Few would question that, although there are fewer unemployed people residing in marginalised rural locations, these are nevertheless highly disadvantaged areas due to the loss of their working-age population cohorts. Many serious problems arise from this fact: high economic dependency rates (i.e. a larger number of people depend on a single income), a loss of the very people who are most important for investing in the locality (both in terms of material investments and social/cultural activities), a decline in reproductive capacity to maintain population levels and a loss of attractiveness to industries and services, to mention only some of the medium to long-term effects.

It is informative to consider some of the findings emanating from international research on rural-urban migration in this context. Todaro (1977) in his path-breaking research on this phenomenon in the Third World, pointed out that migration rates in excess of urban job opportunity growth rates are not only possible but rational, because the decision to migrate depends on the *expected* rather than *actual* urban-rural wage differentials. Creating additional job opportunities at the urban centres as opposed to the migrants' points of departure can therefore have the negative effect of increasing the rural exodus. The discussion about the desired future population distribution in Ireland is only just re-emerging, but we are assuming in this paper that a further concentration of population in Dublin and the Eastern region is undesirable.

A purely "material" deprivation index would, however, have the inevitable effect of exacerbating the "Todaro effect".

We therefore see it necessary to include variables beyond measures of *material deprivation* (e.g. the unemployment rate, levels of car ownership, number of medical card holders or overcrowding) which capture some of the *structural weaknesses* within local communities.

Choice of Variables

A second major decision in constructing an index relates to the question of whether the analysis should focus on actual outcomes or include vulnerable groups. People may belong to a vulnerable group such as the elderly, ethnic minorities or lone parents, without necessarily being deprived, which problematises the use of indicators of the relative magnitude of these groups in deprivation indices. This argument was first put forward by Townsend *et al.* (1988) and has since been followed by a number of other researchers. The authors of the UK 1991 Deprivation Index (DoE, 1995), for example, decided that their index was to be exclusively a measure of outcomes.

In our view, the distinction between "outcomes" and "vulnerable groups" is somewhat artificial. Whilst it is true that not all elderly people or people belonging to an ethnic minority are deprived, the same applies to those who are unemployed or who live in rented accommodation or even those who do not own a car. What is common to all of these indicators, however, is that they *correlate* strongly with poverty. This, of cause, needs to be validated through separate studies: in the Irish case belonging to an ethnic minority would not be a good indicator, whilst the incidence of lone parenthood and small farming certainly are. How spurious the distinction between "material" and other indicators is may best be demonstrated by using the concept of social class. Whilst Townsend (*ibid.*) excludes this variable from his index, Carstairs and Morris (1990) include it in their index precisely because they see low social class as indicating earnings at the lower end of the income scale. We would add to this consideration that social class and educational achievement are the two most important indicators of the life-long opportunities open to any individual, and they constitute important factors in the reproduction of poverty over time. We therefore believe that they should be central to any broad-based deprivation index.

A further controversy surrounds the question of whether variables may be included which measure affluence or the abundance of desirable attributes, possessions and opportunities rather than deprivation or the absence of essential items. The answer to this question cannot be given without reference to the purpose for which the index is being constructed. If we aim to construct an index of broad structural and cumulative disadvantage, as is the case with the present index, then we will be interested in the continuous spectrum from the most affluent to the most deprived areas. Moreover, there is no reason why we should not use low scores on variables which measure affluence (e.g. proportion of the labour force with a third-level education) as indicators of deprivation, if this is consistent with our theory. It may be useful in this context to look at the variables included in the most widely-used deprivation indices in Britain and Ireland (Table 2.1).

Table 2.1. Comparison of the Most Widely-used Deprivation Indices in Britain and Ireland

Variable Census	Townsend	Carstairs	Jarman	DoE 1981	DoE 1991	NI 1991	Haase 1991
Unemployment	✓	✓	✓	✓	✓	✓	✓
Males in part-time employment						✓	
Small farming							✓
Overcrowding	✓	✓	✓	✓	✓	✓	✓
Not owner-occupied/ LA housing	✓						✓
No car access	✓	✓			✓	✓	✓
Lacking amenities				✓	✓		
Children in unsuitable accommodation					✓	✓	
Children in low earner household					✓	✓	
Educational participation					✓	✓	✓
Low social class		✓	✓				✓

Variable								
Single parent			✓				✓	
Under age 5			✓					
Lone pensioners			✓					
Pensioners with no central heating						✓		
Age dependency							✓	
One-year immigrants			✓					
Ethnic minorities			✓	✓				
Vacant dwellings				✓				
Households with no bath, shower or WC						✓		
Properties without public sewerage						✓		
Permanent sickness					✓	✓		
Large household				✓				
Non-Census Variables								
Mortality					✓			
Long-term unemployment					✓	✓		
Income support					✓			
Home insurance					✓			
Low education					✓			
Derelict land					✓			
Rateable Values						✓		

THE METHOD OF ANALYSIS

Over the past two decades a number of different multivariate deprivation indices have been developed, the most well-known being the Jarman Index of Underprivileged areas (1983), the DoE 1981 Index of Deprivation (DoE, 1983), the Scottish Development Department Index (Duguid and Grant, 1983), the Townsend Index (Townsend, Phillimore and Beattie, 1988), the Carstairs Index (Carstairs and Morris, 1990), and, most recently, the UK Index of

Local Conditions (DoE, 1993) and the Northern Ireland Index of Relative Deprivation (Robson et al., 1994).

Whilst all of these indices aim to rank areas in terms of their level of deprivation defined by a range of census variables, the methodology used varies. The Jarman and 1981 DoE Indices first standardise each of the utilised variables and then use a simple additive calculation to derive an overall index. Townsend and Carstairs and Morris use a principal components analysis which identifies the series of components which successively account for the greatest share of variance in the indicator variables. The main advantage of this approach over a simple additive one is that it avoids the "double counting" which occurs when the observed variables are highly correlated. Robson, in his 1991 Deprivation Indices for the UK and Northern Ireland, has since called into question this approach and argued for a chi-square analysis, on the basis that a purely relative measure of deprivation may attach too much importance to calculations based on very small numbers.

Whilst criticism of the simple additive approach is widespread within the research community, it is less certain whether the chi-square approach used by Robson is an improvement over the more common factor-analytical one. Whilst the Robson Index appears more robust in less populous areas, in our opinion it shares the major shortcoming of "double counting" which applied to earlier approaches. As the present index is constructed with a view to aggregating individual DEDs into larger population areas for policy intervention, we can effectively ignore the instability of the results for sparsely populous areas, and the preferred approach is factor analysis.

Once a factor-analytical approach has been selected, we must decide how many variables and factors to include in the analysis and we must consider how the factors can best be interpreted. On the issue of the number of factors, all of the indices referred to above based on a factor-analytical approach use the first component of a principal components analysis as a measure of deprivation. We do not agree with this practice and stressed from the outset that deprivation is a multidimensional concept. Whilst all of our observed variables correlate highly with some other variables, it is argued that they are best represented by a small number of distinct dimensions. This can be demonstrated most clearly in the case of the unemployment rate, which correlates strongly with other deprivation proxies of an urban character, but correlates much more weakly with the more "rural" indicators (small farming etc.). It should be noted here that most existing deprivation indices have originated from England and Scotland. For highly industrialised countries, the use of just the first principal component may be justified. To apply the same approach to Ireland, however, which is considerably more rural in character and where many rural areas have experienced decades of practically uninterrupted emigration, seems to us to be missing a major point.

Before discussing the interpretation of factors, we should consider what kind of rotation should be applied after initial extraction. The default rotation, and the one most commonly used, is varimax rotation, which maintains an orthogonal structure of the individual factors; the factors are calculated so that they remain completely independent of one another. In our view, this is an unwarranted constraint in the deprivation context, as it is unlikely that the underlying forces that create and reproduce deprivation are themselves uncorrelated. By allowing oblique rotation (for example using the oblimin procedure), the structure matrix will be chosen in such a way that the observed variables will load more cleanly on one or another of the factors, thereby facilitating a more straightforward interpretation of the factors themselves. It should be noted that the choice of orthogonal or oblique rotation has no impact on the overall variance that is explained by the factors.

Despite having opted for a factor-analytical approach in the design of the Irish Deprivation Index, we would like to draw attention to another concern which makes the use of this approach problematic and which has not been addressed by any of the existing indices. This is the inherently unstable structure resulting from each analysis. As equivalent analyses are undertaken for different regions, or for the same region at a different level of geographical aggregation, or for the same region and the same spatial units, but at a different point in time, the resulting structure matrix changes. This makes it impossible to compare results between two regions, or between different points in time. The reason for this is that the approach used here (principal components analysis) is a type of *exploratory* factor analysis where the relationships between the components and the variables depend on the dataset used. We intend to use confirmatory techniques in future research in an attempt to overcome these limitations.

Construction of a Single Index

Any multivariate index has to deal with the problem of how to combine scores on a number of variables into an overall deprivation score. As already noted above, the additive and chi-square approaches are both vulnerable to the charge of "double counting". However, use of the first principal component may neglect potentially important dimensions of deprivation which, as we pointed out above, seem to be particularly important in the Irish context. Unfortunately, there is no objective or "scientific" way to determine the "correct" number of factors or components to include in a given analysis. Furthermore, there is no single correct way of combining the scores for multiple components into an overall measure of deprivation. However, we can argue that deprivation is driven by a series of different processes, whereby some areas are disadvantaged due to prevailing labour market conditions and others on account of their underlying demographic composition. Given this fact, we feel justified in selecting more than one component, estimating scores for those components and then summing the resulting scores to arrive at an overall measure. The justification for this approach comes from theory, as where we argued that persistent emigration in itself is the result of long-term labour market weaknesses which we cannot capture in terms of the local unemployment rate.

Validation

The external validation of deprivation indices poses a major problem. At the outset of this chapter, we conceptualised deprivation as an *absence of essential or desirable attributes, possessions and opportunities which are considered no more than the minimum by society.* The individual variables used in the construction of an index are chosen as potential indicators of these dimensions (e.g. a balanced population profile, car ownership and social class as a measure of overall opportunities). The multi-dimensional structure of deprivation means that no single variable will be adequate for external validation. As many of the deprivation indices were originally designed in the context of health needs, one common criterion of validation is the correlation with the standardised mortality rate (SMR). SMRs measure the number of premature deaths (all causes) per 1,000 population, after adjusting for the specific age profile of an area. Unfortunately, in Ireland, SMRs have only recently been calculated at DED-level for the Dublin area, and do not yet exist at DED level for the country as a whole. SMRs are generally highly correlated with the existing deprivation indices, although such correlations are typically almost twice as high in urban areas than in rural ones. In the absence of SMRs for the country as a whole, little can be gained by validating an index for Dublin only, particularly as one of the specific features of the Index presented here for Ireland is its measure of rural deprivation.

Alternatively, we could use administrative data for the take-up of social welfare benefits. Again, such data are not available for Ireland, but do exist in Northern Ireland. Work undertaken subsequent to the delivery of this paper at the GSI conference showed that, for Northern Ireland, the Haase Index of Deprivation had higher correlations than either the Robson or Jarman Indices with all currently existing ward-level benefit take-up data (Disability Living Allowance (0.60), Income Support (0.81), Unemployment Benefit (0.74), Family Credit (0.69)). It should further be noted that the benefit take-up variables were correlated more than twice as strongly with the three deprivation indices as the SMR.

SOURCES OF POVERTY AND SOCIAL EXCLUSION

In this section we will discuss the relationship between poverty and social exclusion, on the one hand, and a series of factors which have been highlighted as potential causes of these. It is not enough to demonstrate a statistical relationship between poverty and social indicators, for we must also provide a substantive discussion of how the variables used in our deprivation index capture some of the key determinants of poverty. The most important causes discussed in the literature are unemployment, on-farm underemployment, isolation, lone parenthood, low social class, educational achievement and type of housing. It should also be stressed that gender and age can interact with these factors in complex ways to generate forms of poverty and social exclusion which impact, for example, on women or old people in a particular way. The overlapping of several of these causes (unemployment, age profile, educational achievement, for example) can give rise to very severe forms of deprivation in local areas (multiple or cumulative deprivation) which can, in turn, have an independent effect on the opportunities available to the people living in these areas.

Unemployment

Clearly, the most crucial determinant of poverty in Ireland is, as in most other developed countries, the failure to find paid employment. In the cities, deprived areas tend to coincide with well-known unemployment blackspots. In rural areas, where unemployment rates tend to be slightly below those found in urban areas, the spatial link between unemployment and deprivation is weaker, firstly because unemployment is less concentrated and secondly because it may be disguised. Nevertheless, the unemployed represent the largest proportion of those who are economically poor and live in rural areas.

On-farm Underemployment

In rural areas, unemployment, and hence the risk of poverty, may be further exacerbated by hidden unemployment in the form of on-farm underemployment. It is by now well established that the vast majority of farms below 50 acres do not provide a sufficient farming income for the owner occupier; these small farmers generally depend on some form of state transfer payment or an alternative source of income (Moss *et al.*, 1991). Poverty, in its economic sense, must therefore be assumed to be widespread amongst small farmers. However, this may not result in the same level of social exclusion. As the ownership of land confers a certain status upon the holder within the rural community, holding onto otherwise unviable holdings (in economic terms) may be seen as a strategy to avoid social exclusion.

Women in Economically Poor Farming Households

Whilst holding onto economically unviable farming enterprises may confer some degree of "social inclusion" upon the owner, this effect varies considerably *within* the farming household. Village life still remains predominantly male in character, as travel, employment mobility, social contact and, most tellingly, use of village space, is consistently curtailed for women. Whereas many men have access to private transport either to conduct business outside the village or to work in neighbouring towns and villages, this is largely unavailable to women. Poor public and private transport provision and the constraints of childcare make it difficult, if not impossible, for women to work outside the farm or local village. It is therefore the case that women, particularly when engaged in small farming, are generally exposed to a greater risk of social exclusion than men.

Elderly People Living on their Own
Persistent emigration over the past decades has resulted, in many rural areas, in distinctive population characteristics. As emigration is concentrated amongst the core working age cohorts, this leaves behind increasing proportions of economically dependent age cohorts, i.e. the very young and the elderly. In many remote rural locations, this has resulted in population imbalances characterised by the extreme thinning out of the main agents of economic activity, and by high proportions of elderly people living alone. These areas typically also have high proportions of unmarried men over 45 years of age indicating the delay, if not abandonment, of family formation.

Lone Parents
Many lone parent families find themselves entirely, or almost entirely, dependent on social welfare payments due to the difficulty of finding work which is sufficiently remunerative to fund child care as well as the difficulty of organising child care in itself. Where lone parents manage to overcome these obstacles, the fact that substantial costs must still be met from a single income often means that these families face substantial hardship. Therefore, the situation of lone parent families is determined by both the level of social welfare payments and the organisation of work and child care in contemporary society. The inadequate level of social welfare is highlighted by the fact that such payments still leave the recipient in a situation of poverty as defined by any of the commonly used relative income thresholds.

Increasing numbers of families find that a single income is inadequate to meet their needs; for lone parent families this means increasing hardship and reliance on the state. Research conducted during the late 1980s and reported in Nolan and Callan (1994) indicates that the presence of young children, in itself, is associated with a greater likelihood of poverty at household level, presumably due to the resources required to provide for them. This life-cycle effect is exacerbated for lone parents because many of the coping mechanisms available to two-parent families (increased overtime work, a second job, organising work to facilitate parental child care etc.) are simply unavailable to lone parents.

Low Social Class
The dominant theoretical approaches to the subject of class emphasise that social class is not an individual attribute, but a social relationship which expresses the degree of control an individual can exercise over important resources in society. At the lower end of the social class scale, the concept provides a unified way of thinking about the situation of people with different types of job or with no job at all, and for this reason it is crucial to all discussions of poverty. What justifies the inclusion of a heterogeneous group of people in a single social class category is the fact that they have a similar relationship to productive resources. This social relationship is associated with very different opportunities in labour market terms and in relation to lifestyle and life possibilities. Thus a low social class composition does not merely indicate that people in a given area have "earnings towards the lower end of the scale", but suggests a structural weakness which may manifest itself in below-average educational attainments, employment instability, unemployment and a generally weak labour market situation.

As social class cannot be observed directly and is broader than either "occupation" or "status", we need to use a range of variables to measure it, including property, educational and occupational variables. The census variable "social class" captures part of this picture, as it assigns individuals to class categories depending on both their occupation (higher or lower professional, non-manual, semi- or unskilled manual) and their possible land holdings. The "higher professional" social class includes farmers with 200 acres of land and more; those working between 100 and 199 acres are allocated to the "lower professional" class. The next class, "other non-manual", includes farmers with between 50 and 99 acres, and those with 30-49 acres are described as "skilled manual". Finally, the "semi-skilled manual" category includes "small farmers" — those with less than 30 acres of land. This variable is, therefore, a pretty good guide to the social class composition of neighbourhoods throughout Ireland. It is, however, flawed because the divisions between its categories are based on occupational rather than social class criteria, and because a significant proportion of people do not fall into any of its categories.

In relation to the first point, it has been argued that the distinctions between manual and non-manual employees and according to skills are not the most important divisions within the workforce. In relation to categorisation, the percentage of people whose social class position is "uncategorised" can reach very significant levels — for example, this figure exceeds 40 per cent in the most deprived parts of Dublin, Waterford and Limerick cities — casting substantial doubt on the accuracy of the picture presented by the remaining categories. As this seventh "unclassified" category includes people who have never been in paid employment, it seems likely that the semi- and unskilled manual categories will understate the extent of disadvantage in these areas.

Low Educational Achievement
Closely related to the social class composition of an area are its levels of educational attainment, measured either in terms of school leaving age or educational qualifications obtained. The close relationship between class and education have encouraged some social scientists to argue that class background conditions young people's experience of the school environment and their likelihood of academic success (Willis, 1986). Thus, as academic "credentials" become increasingly important in gaining access to certain forms of employment, educational attainment becomes one of the mechanisms by which class inequality is reproduced from one generation to the next. In concrete terms, unemployment and low-paid employment go hand in hand with early school leaving and lack of educational attainment. In order to understand this empirical association, it should be noted that for many disadvan-

taged school students, the perceived likelihood of unemployment feeds into disillusionment with the educational system and vice versa: both structures form part of a system of inequality and poverty which is highly resistant to change.

Local Authority Housing
Most poor people do not live in local authority housing estates, and even within such estates it is not necessarily true that the majority of inhabitants are economically poor. However, it is also true that these estates have become the locations of major clusters of unemployed people and of those who have access only to low-skilled and poorly-paid employment. In some of these estates, unemployment levels are so high that the lack of role models for the younger generation leads to high incidences of early school-leaving and the inevitable reproduction of social disadvantage from one generation to the next. The unfavourable social conditions prevailing in the larger urban local authority housing estates are well documented. Whilst smaller local authority housing estates in rural locations may not attract the same public attention as their urban counterparts, social deprivation in these estates has an added dimension: people living in these estates are excluded from society in that they do not own land and therefore have little status within the rural community. Nor do they form part of village life, from which they are systematically excluded by geographical location (most of these estates are situated on the outskirts of villages and towns). Poor housing conditions often exacerbate this situation, and these housing estates must be considered as locations of multiple or cumulative disadvantage.

THE IRISH INDEX OF DEPRIVATION

On the basis of the considerations outlined in the previous section, the analysis presented here uses thirteen indicators that have been proven to be strongly associated — directly or inversely — with the presence of deprivation:

1. The age dependency rate

2. The proportion of lone parents

3. The unemployment rate

4. The percentage of those at work engaged in small farming (under 30 acres)

5. The proportion of households with two or more cars

6. The percentage of the population in the combined higher and lower professional classes

7. The percentage of the population in the unskilled manual class

8. The percentage of economically active persons with third-level education

9. The percentage of the adult population leaving school at 15 years or below

10. The percentage of the adult population leaving school at 20 years or above

11. The proportion of permanent private households which are local authority rented

12. The proportion of permanent private households which are owner occupied

13. The average number of rooms per person.

Principal components analysis results in the identification of three dimensions which account for 73 per cent of the total variance observed (Table 2.2). The first factor, accounting for 39.7 per cent of variance, appears to measure *social class*; the other two factors (accounting for 24.1 and 8.7 per cent respectively) are related to the distinct features of *urban deprivation* and *rural deprivation*. Large proportions of people in local authority housing, lone parents and unemployment are predominantly indicators of deprivation in urban settings, whilst the proportion of small farmers and the age dependency rate are both distinctly rural. The proportion of households with two or more cars loads almost equally onto both the urban and rural components (obviously with an inverse sign).

Table 2.2. *Structure Matrix of Deprivation Factors Based on Multivariate Analysis of 1991 Small Area Population Statistics*

Variable	Factor 1 Social Class	Factor 2 Urban Deprivation	Factor 3 Rural Deprivation
Labour force with 3rd level education	.90		
Leaving school at 20 or over	.89		
Rooms per person	.80		
Leaving school at 15 or under	-.77		.66
Higher and lower professional class	.73		-.62
Unskilled manual class	-.63		
Households owner occupied		-.91	
Households local authority rented		.85	
Lone parents		.78	
Unemployed		.77	
Proportion of small farming			.84
Households with two or more cars		-.54	-.60
Age dependency rate			.56

Coefficients less than 0.5 omitted.

Source: Haase, 1995.

The clear emergence of the overriding influence of social class together with two distinct but clear indicators measuring urban and rural deprivation strongly concurs with the theoretical ideas present at the outset of the analysis. After assigning individual scores for the three factors to each DED, these were then aggregated to derive a single indicator of the overall degree of deprivation prevailing in each. Figure 2.1 shows the resulting distribution of relative affluence and deprivation throughout Ireland.

Figure 2.1: Affluence and Deprivation in Ireland, 1991

THE GEOGRAPHICAL DISTRIBUTION OF POVERTY

The distribution of deprivation, as measured by this procedure, clearly supports Nolan and Callan's (1994) assertion that it is a spatially pervasive phenomenon which affects almost every part of the country. However, there are differences in the degree to which disadvantage is clustered in particular areas, both urban and rural. In Dublin, the analysis fully confirms the known areas of deprivation. These are the North and South Inner City, Coolock, Ballymun, Cabra, Finglas, parts of Blanchardstown, Rialto, Kilmainham, Ballyfermot, Cherry Orchard, Clondalkin, Kimmage, Crumlin, Walkinstown, West Tallaght, and pockets in Dun Laoghaire. In the urban areas outside Dublin the main clusters of deprivation include parts of the four County Boroughs of Galway, Limerick, Cork and Waterford and the towns of Drogheda, Dundalk, Sligo, Wexford, Bray and Kilkenny. In rural Ireland, disadvantage is most prevalent in counties Donegal and Mayo, but also extensive in the border counties of Leitrim, Cavan, and Monaghan, as well as in Roscommon. Further significant pockets include North Kerry and parts of County Clare.

The second feature that is apparent from this map, and which could not have been deduced from either a county-level analysis or the spatial categories applied in the ESRI income study, is the extent to which the fate of rural Ireland seems to be determined by urban factors. The interpenetration of the countryside by the urban shadow of the major centres of population could not be more striking. A pattern emerges first of relative affluence and then increasing disadvantage in concentric circles around the main centres of population. This is particularly apparent in the case of Dublin and the cities of Cork, Galway, Limerick and Waterford, but equally applies to Sligo, Dundalk, Drogheda, Cavan, Monaghan, Athlone, Ennis, Kilkenny, Cashel, New Ross and Wexford. Indeed, it is probably correct to say that this effect can be observed in almost every town throughout the country. In each case, one can observe a ring of affluence stretching from the outer fringes of suburban developments into the adjoining rural space. Therefore, the degree to which more extensive rural areas become disadvantaged appears to be a function primarily of the relative density and size of the urban centres located in its proximity.

This effect most likely results from a number of underlying developments. Firstly, an increasing number of people are commuting into the urban area due to the improvements in road and rail networks in recent years, which make it practical to live well outside the suburban ring and still be no more than an hour from the centre of the city or town. Many of these will be people who have benefited from the opportunity to buy property at a much lower cost than in the city and who may have benefited from greenbelt provisions in the planning legislation to preserve some of the amenity values of their property in the face of future development. It includes a number of satellite communities that have attracted a degree of industry and employment precisely because they are located near an urban centre, an airport and the range of cultural amenities that the city or town offers. It also includes agricultural landholders who, if their farms are large, have benefited from the opportunities for the intensive commercialisation of agriculture provided by the nearby urban market or, if they are too small to be economically viable, have benefited from opportunities for off-farm economic activity in the city or town. In any case, for many landholders located in the proximity of urban centres, the proceeds from selling parcels of land with planning permission for residential use has often provided exceptional windfall incomes.

This process of urban extension is likely to continue. Its corollary, the depopulation that is occurring both in the central areas of the cities/towns, and the more peripheral rural areas, coincides with the significant areas of deprivation shown in the map. What these patterns indicate very clearly is that one of the forces at work is the effect of land values in determining the life chances of the population. As the expansion of the urban fringe has continued, it has tended to exacerbate the situation of the propertyless and landless rural and urban populations, as they have been forced to compete in a market inflated by urban interests and in a social structure and culture that is increasingly defined by urban interests.

REFERENCES

Carstairs, V. and Morris, R (1990) Deprivation and Health in Scotland, *Health Bulletin*, 48, 162-174.

Curtin, C., Haase, T. and Tovey, T. (eds) (1996) *Poverty in Rural Ireland — A Political Economy Perspective*, Oak Tree Press and Combat Poverty Agency, Dublin.

Duguid, G. and Grant, R. (1983) *Areas of Special Need in Scotland*, Central Research Unit, Scottish Office, Edinburgh.

Great Britain — Department of the Environment (1995) *1991 Deprivation Index: A Review of approaches and a Matrix of Results*, HMSO, London.

Haase, T. (1993) *Identifying Prospective Areas for Inclusion in the Local Development Programme*, Briefing Paper to the Combat Poverty Agency, Dublin.

Haase, T. (1995) *The Designation of Disadvantaged Areas in the Local Development Programme*, Report to Area Development Management Ltd. to facilitate the designation of areas under the Operational Programme for Local Urban and Rural Development, 1995-1999.

Jarman, B. (1983) Identification of underprivileged areas, *British Medical Journal*, 286, 1705-1709.

Moss J *et al.* (1991) *Study of Farm Incomes in Northern Ireland and the Republic of Ireland*, Co-Operation North, Belfast.

Nolan, B. and Callan, T. (eds) (1994) *Poverty & Policy in Ireland*, Gill and Macmillan, Dublin.

Robson, B., Bradford, M. and Deas, I. (1994) *Relative Deprivation in Northern Ireland*, Policy Planning and Research Unit, Occasional Paper No. 28, HMSO, Belfast.

Rottman, D. (1994) *Income Distribution within Irish Households*, Combat Poverty Agency, Dublin.

Todaro, M.P. (1977) *Economic Development in the Third World*, Longman, New York.

Townsend, P., Phillimore, P. and Beattie, A. (1988) *Health and Deprivation: Inequality and the North*, Croom Helm, London.

Williams, J. (1993) *Spatial Variations in Deprivation Surrogates — A Preliminary Analysis*, Report to the Combat Poverty Agency, ESRI, Dublin.

Williams, J. (1995) *Spatial Aspects of Poverty and Disadvantage*, Paper prepared for the National Economic and Social Forum, ESRI, Dublin.

3

SPATIAL ASPECTS OF POVERTY AND DEPRIVATION IN IRELAND

Brian Nolan, Christopher T. Whelan and James Williams
Economic and Social Research Institute, Dublin

INTRODUCTION

The purpose of this chapter[1] is to present some details on spatial aspects of poverty and deprivation in the Ireland of the early 1990s. Much has been written on various aspects of poverty over the last 10 years. One issue which appears to have avoided the focus of researchers' attention has been the question of the geography of poverty and the role of geography in determining poverty risk and incidence levels. The relationship between geography, poverty and deprivation are clearly of importance to the policy-maker in designing strategies aimed at addressing the issues involved. For example, evidence of a substantial concentration of poverty in particular areas (urban or otherwise) or indeed in particular types of areas would clearly be of importance in devising policy prescriptions. Area-based responses to poverty have, in recent years, assumed an increasingly important role in efforts to address the problem. Indeed, policy and strategy in this field seem to have outpaced the development of a theoretical framework. Relatively little is understood about the relationship between space and poverty. The intention of this chapter is to begin an examination of some aspects of this relationship.

This chapter is divided into seven sections. In section 2 we consider some basic definitions of poverty and deprivation; in section 3, we briefly discuss data sources used in this chapter. Section 4 examines variations in poverty risk and incidence levels according to a classification of areas by size and tenure type. Section 5 looks at poverty levels by planning region. Section 6 discusses variations in deprivation surrogates using data from the Small Area Population Statistics and, finally, section 7 provides a summary and conclusion.

POVERTY AND DEPRIVATION — SOME DEFINITIONS

Poverty is a very emotive word and views differ substantially as to what it means or implies. In Third World countries poverty is usually accepted as a condition in which people are struggling to survive, struggling to provide reasonable shelter, to avoid starvation or to meet some minimum standard of nutrition. In such a society there is relatively little dispute as to the meaning of poverty. Physical survival is clearly central to any definition we might choose to adopt. In the developed societies of the first world, however, there is much more debate surrounding the meaning of poverty and deprivation. Although few people die of starvation in Ireland today, this does not mean that there is little or no poverty here. In the developed world, poverty can be seen as being defined by the needs of society as a whole. Accordingly, what is taken as constituting poverty is influenced by the general socio-economic conditions of the population in question. Because it is linked to the conditions of the society under examination, the definition of an "adequate" standard of living varies across countries and, indeed, varies over time within the same country in line with changes in social norms and values. Sixty years ago, for example, not many households in Ireland had hot and cold running water. Today most households have running water and it is regarded as a necessity. In this respect, poverty is very much a relative concept. As far back as the 18th century Adam Smith noted in *The Wealth of Nations* that:

> . . . necessities include not only the commodities which are indispensably necessary for the support of life, but whatever the custom of the country renders it indecent for creditable people, even of the lowest orders, to be without.

This vein of thinking, which views "needs" and "necessities" as being inevitably determined by the society in question, has been continued up to the present day in most of the contemporary work on poverty. Much of the recent work on poverty traces its lineage from Peter Townsend's work on poverty in the UK when he noted that:

> Individuals, families and groups in the population can be said to be in poverty when they lack the resources to obtain the type of diet, participate in the activities and have the living conditions and amenities which are customary, or at least widely encouraged, or approved, in the societies to which they belong. Their resources are so seriously below those commanded by the average individual or family that they are, in effect, excluded from ordinary living patterns, customs and activities (Townsend 1979, 31).

Poverty, therefore, is about relative exclusion due to a lack of resources. One way of measuring this involves the use of so-called relative income poverty lines. These lines measure poverty relative to the standard of living in society as a whole. The poverty line can be drawn at any level in the income distribution. For example, the 50 per cent line takes the poverty threshold as 50 per cent of average household income. To account for differences in household size, structure and composition, one uses a set of equivalence weights to derive a measure of what is referred to as equivalised household income or household income per adult equivalent member. The 60 per cent poverty line, therefore, would be located at a point representing 60 per cent of average equivalent household income. Relative income poverty lines have become something of an industry standard in work in this field.[2] By definition, they relate specifically to measurable income (or lack of it).

A further aspect of well-being is consumption, style of living and deprivation. The work on deprivation indicators was initiated by Townsend (1979) and developed by Mack and Lansley (1985). More recent international work in this area has been undertaken by Townsend and Gordon (1989), Mayer and Jencks (1988) and Muffels and Vriens (1991). Work on non-monetary measures of poverty was first carried out in Ireland on the basis of a survey of lifestyles and living standards conducted by the ESRI in 1987 and subsequently reported in, among others, Callan *et al.*, (1988, 1989); Callan and Nolan (eds) (1994); Whelan *et al.*, (1991);

Callan, Nolan and Whelan (1993); Whelan (1994) and Nolan and Whelan (1996). These studies derived a number of indices of deprivation based on the possession or otherwise of a list of 24 items. When constructing the indices, particular emphasis was placed on situations where the absence of an item was due to financial constraints experienced by the household. Using factor analysis to combine the 24 items in question, three main consumption indicators were derived, viz. a basic or primary consumption indicator; a housing indicator and an indicator of secondary deprivation.[3] The primary consumption indicator was derived from eight basic items relating to a lack of food; clothes; adequate heating; and a high level of household debt. Households were assigned a score of "1" for each item which they lacked.

By combining the relative income poverty measures with the deprivation indicators (especially with the indicator of basic or primary deprivation) one can derive a measure of exclusion from the *combined* societal norms of both *income* and *consumption*. In other words, it gives a measure of exclusion based on a lack of resources in consumption which is enforced on a household due to financial constraints. To strike the combined relative income and basic deprivation poverty line one defines poverty as being below (for example) the 60 per cent average income threshold and also experiencing a score of one or more on the basic or primary deprivation. Throughout the remainder of this chapter we will focus on trends in poverty and deprivation using the combined 60 per cent relative income/basic deprivation line.

THE DATA

This chapter presents findings based on two data sources. The first is a nationally representative sample of just over 4,000 households carried out by the ESRI in 1994.[4] Among a host of other information, it collected details on household composition; income; general financial and economic well-being; receipt of social welfare and other transfers; issues relating to health, educational attainment, labour market experience, etc.

A national random sample of 4,000 households (along with detailed interviews with the 10,000 individuals who lived in them) is extremely large by the standards of social science data-sets.

Despite its size, however, one is still constrained in the degree to which one can disaggregate the results by any given characteristics, including geographical area. If one tries to produce results for recognisable geographical units (for example counties) one quickly runs out of cases in individual cells. What one can do, however, is to carry out analyses according to descriptive characteristics of type *of area* such as rural or urban location; towns; cities etc. This allows us to investigate variations in poverty *rates* by "area-type" as well as the extent to which poor households are concentrated within regions or types of regions across the country. In addition, and notwithstanding the constraints imposed by sample size, some analysis of poverty rate and incidence levels can be undertaken at the scale of the planning region.

The analysis from the cross-sectional data is complemented with some analysis of the 1991 Small Area Population Statistics data (SAPS). Because the SAPS provides us with no details on individual households, one cannot use them to derive a direct estimate of the number of households in poverty. What is possible, however, is to look at geographical variations in the prevalence of the characteristics which are known to be correlated with poverty. A complete analysis of these surrogates would allow us to investigate the geography of deprivation in Ireland. In the space allowed in this chapter we use the SAPS data only to briefly and tentatively examine some issues associated with the *rate* of deprivation and, more importantly, its *concentration* across geographical units.

POVERTY, AREA-TYPE AND STATUS OF TENURE — SOME CROSS-SECTIONAL RESULTS

Table 3.1 presents details on the risk and incidence of poverty by *type* of area in which households are located. The figures on the *risk* of poverty give the probability of a household in the specified type of area being in poverty whereas the *incidence* figures provide a breakdown of households which are below the poverty line according to the area-type classification. The area types used in the classification are as follows: (a) open country; (b) village or town with a population of less than 3,000 persons; (c) a town with a population of 3,000 or more; (d) the cities of Cork, Limerick,

Waterford or Galway and (e) Dublin city and county. The table presents information based on the combined 60 per cent relative income line plus basic deprivation. From this we can see that the lowest risk levels are experienced by households in open country (8.3 per cent) followed by those in Dublin city and county (14 per cent). These compare with an aggregate average level of 13.7 per cent for the county as a whole. According to this poverty threshold, highest risk levels are experienced by households in the four county boroughs as well as those located in small towns or villages (20.6 per cent).[5] These initial figures suggest, therefore, that high levels of poverty risk are not the sole preserve of large urban areas.

Table 3.1: Risk and Incidence of Poverty by Type of Area, 1994

Type of Area	Sixty Per Cent Income Line + Basic Deprivation		Percentage of All Households
	Risk	Incidence	
Open country	8.3	20.0	33.1
Village/Town < 3,000	20.6	15.8	10.5
Town > 3,000	16.5	21.8	18.1
Four county boroughs*	19.4	12.4	8.7
Dublin city and county	14.0	30.1	29.5
Total	13.7	100.0	100.0

* Waterford, Galway, Limerick and Cork County Boroughs.

Table 3.1 also provides details on the incidence or distribution of the households which are in poverty according to the area-type classification. From this one can see, for example, that when using the combined income and deprivation line a total of 12.4 per cent of households which are in poverty are located in the four county boroughs outside Dublin; 16 per cent in small towns/villages; 20 per cent in open country; 22 per cent in medium-sized towns and 30 per cent in Dublin city and county. An indication of the degree of concentration or otherwise of poverty can be found by comparing the second last column (showing poverty incidence levels) with the last column which shows the percentage of all households in each of the area types in question. From such a comparison one can see that the incidence of poverty in Dublin (30 per cent) is very much in line with the percentage of all households (29.5 per cent) in the city and county region. Using this criterion it is clear that the highest level of over-concentration is in the four county boroughs outside Dublin. These contain 12.4 per cent of households below the line compared with only 8.7 per cent of all households. Medium sized towns and small towns/villages each contain an over-concentration of poor households. There is clearly an under-representation of poor households in open country when we use the combined poverty line, with these areas accounting for 20 per cent of poor households and 33 per cent of all households. The story told by these figures would suggest, therefore, that although there are certainly examples of over- or under-representation of poor households across the area-types used in the table, in general, poor households are not exclusively concentrated in one or two of the area types discussed. It is particularly clear, for example, that they are not wholly concentrated in urban areas.

This analysis of poverty by area type can be complemented with a consideration of poverty risk and incidence levels according to the nature of household tenure. This allows us to consider variation in poverty rates between the public and private housing sectors — which tend to be spatially segregated — and, specifically, to see whether or not households in poverty are concentrated in the public housing areas. This information is presented in Table 3.2. From this we can see that the risk of poverty is highest for households in the local authority rented sector. Just under 47 per cent of households in this sector were in poverty in 1994 using the combined relative income/deprivation thresholds compared with, for example, the aggregate average figure of 13.7 per cent. The second highest risk is for the local authority tenant — purchase sector (23.1 per cent). The risk is lowest for the private sector "owned with a mortgage" (7.2 per cent).

If we change the focus from *risk to incidence* classified by nature of tenure we can see that the high risk of poverty facing those in the local authority rented sector is reflected in a substantial concentration of poor households in public housing. Using the combined 60 per cent relative income and basic deprivation line as the poverty benchmark, we can see that 40 per cent of poor households are in the local authority rented sector, even though only 12 per cent of all households are classified as being in that sector of the housing market. If one includes households which are on a local authority tenant-purchase scheme we can see that just over half of all households in poverty are in local authority housing. This compares with 18 per cent of all households in total.

Table 3.2: Risk and Incidence of Poverty by Tenure Type, 1994

Type of Area	Sixty Per Cent Income Line + Basic Deprivation		Percentage of All Households
	Risk	Incidence	
Owned outright	8.0	24.5	42.0
Owned with mortgage	7.2	16.7	32.0
L.A. tenant purchase	23.1	10.7	6.3
L.A. rented	46.9	39.9	11.7
Other rented	14.2	8.3	8.0
Total	13.7	100.0	100.0

The above average incidence of poor households in the public housing market is clearly counterbalanced by the lower than average incidence in the private sector. The incidence of households in poverty in the private rental sector (just over 8 per cent) is exactly as one would expect given their distribution in the general population of households as a whole. The incidence among households owned outright is 24 per cent and among those owned with a mortgage is 17 per cent. These incidence rates compare with population figures of 42 and 32 per cent of all households respectively.

In Table 3.3 we present details on risk and incidence levels for households cross-classified by area-type and status of tenure. This table shows risk and incidence levels in (a) the local authority sector and (b) other sectors of the housing market classified by area-type. From the table one can see that there is clearly a substantial differential between the poverty risk among local authority households on the one hand and the private housing sector on the other across all area-types. The lowest level of disparity is in areas of "Open Country". For local authority households in "Open Country" the risk of poverty is 14 per cent. This compares with 8 per cent for their counterparts in the private housing sector in similar areas. This represents a disparity in poverty risk between local authority and other households in the region of 1.75:1. The disparity between the two broad tenure types is substantially higher in all other area-types ranging from 2.8:1 in medium sized towns; to 3.5:1 in small towns/villages; to 6.0:1 in the four county boroughs; and reaching a figure of 8.0:1 in Dublin city and county.

Table 3.3: Risk and Incidence of Poverty in 1994 Using Combined 60% Relative Income Plus Basic Deprivation Classified by Area and Tenure Type

Area Type	Risk of Poverty		Incidence of Poor Households		Percentage of All Households	
	Local Authority	Other	Local Authority	Other	Local Authority	Other
Open country	14.0	8.0	1.7	18.3	1.6	31.5
Village/Town < 3,000	43.5	12.5	8.7	7.1	2.7	7.7
Town > 3,000	33.4	11.8	9.6	12.2	3.9	14.2
Four county boroughs*	43.3	7.2	9.3	3.1	3.0	5.8
Dublin city and county	43.2	5.4	21.0	9.0	6.7	22.8
Total	38.5	8.3	49.6	50.4	17.9	81.1

* Waterford, Galway, Limerick and Cork County boroughs.

If one considers the incidence of poverty classified by tenure and area types one can see that just over 30 per cent of households below the poverty line are located in the local authority sector in the combined areas of Dublin city and county along with the other four county boroughs. A further 12 per cent of poor households were located in these areas in households which were not in the local authority housing market. If one compares these poverty incidence figures with the breakdown of all households in the country as a whole we can see that 9.7 per cent of all households were in the local authority sector in this combined area, while a further 29 per cent were in the private sector. In summary, therefore, just over 42 per cent of households which were in poverty in 1994 were in the four county boroughs and Dublin city and county in 1994.

This compares with 38.3 per cent of all households which were located in these areas in that year. One should note, however, that although 30.3 per cent of all households in poverty were located in the local authority housing market in the combined Dublin city and county areas this obviously implies that 70 per cent of households in poverty were outside these areas in the year in question. Once again, this suggests that poverty is not solely concentrated in the large urban areas (which contain 42 per cent of poor households as compared with 38 per cent of all households). There does, however, seem to be a relatively stronger concentration in the urban local authority sector (30.3 per cent of poor households in local authority housing in Dublin and other four county boroughs as compared with 9.7 per cent of all households).

POVERTY RISK AND INCIDENCE BY PLANNING REGIONS

Because the poverty estimates discussed so far are based on a cross-sectional sample survey, we are limited in the degree to which we can disaggregate our results to recognisable, geographic units (such as counties) before we begin running out of cases. We do, however, have a sufficiently large number of cases[6] to consider risk and incidence figures at the level of the Planning Region. Table 3.4 presents the relevant information. From this one can see that, using the combined 60 per cent relative income plus basic deprivation threshold, risk levels range from a figure of 8.7 per cent in the North-East to a figure of 19.7 per cent in the North-West and Donegal region. Clearly the risk figures provides only one side of the story. To get a more complete picture of issues of concentration one must consider incidence rates as well. This information is also provided in Table 3.4. This shows, for example, that 41 per cent of households in poverty were located in the East region, 15 per cent in the South-West region and so on. For comparative purposes the table includes a breakdown by region of the total population of all households in the State. The striking feature of the incidence figures is the degree to which the distribution across the planning regions of households below the poverty line is so similar to that for all households. This clearly suggests that households in poverty are distributed across the regions in the proportions which one would expect on the basis of the distribution of all households in the population. Clearly, given the highly aggregated nature of the planning regions this apparently equitable dispersal of poverty may mask local concentration at a more spatially disaggregated scale. If we are to investigate the existence or otherwise of these local concentrations we must consider other data. To this end we briefly turn to the 1991 Small Area Population statistics to consider spatial variations in two of the more important proxies for poverty and deprivation.

Table 3.4: Risk and Incidence of Poverty by Planning Region, 1994

Planning Region	Sixty Per Cent Income Line + Basic Deprivation		Percentage of All Households
	Risk	*Incidence*	
East	14.7	41.0	38.1
South-West	13.2	15.0	15.5
South-East	14.6	11.1	10.4
North-East	8.7	3.4	5.4
Mid-West	12.6	8.2	8.9
Midlands	12.0	6.4	7.3
West	8.9	4.8	7.4
North West + Donegal	19.7	10.0	6.9
Total	13.7	100.0	100.0

DEPRIVATION SURROGATES FROM THE SAPS

Much of the Irish research[7] into the area of poverty and disadvantage indicates the importance of factors such as social class and employment status in determining poverty. By examining spatial variations in these variables we can, therefore, advance our understanding of the nature of the geography of poverty, deprivation and disadvantage.

We can do this using the 1996 Small Area Population Statistics (SAPS). Although the information contained in the SAPS does not provide us with any details on the social class or employment composition of *households* it does allow us to examine the class and labour force structure of *areas*. By discussing such structures one can infer a substantial amount about spatial variations in both the risk and incidence of poverty and deprivation, particularly as regards its concentration at a geographically disaggregated level.[8]

Social Class

Class Distribution at the National Level
The first proxy of disadvantage and deprivation to be considered is social class. Table 3.5 presents information on the percentage of each county's population which fall into the combined Higher and Lower Professional group and the Unskilled Manual category — the figures in parentheses indicate rank. A total of 25.2 per cent of the total population in the State fell into the Professional classes in 1991 while 10.4 per cent fell into the Unskilled Manual category. At the county level we can see from Table 3.5 that Dublin county had the highest percentage of its population falling into the combined professional category (34.4 per cent). This was followed by Wicklow (29.4 per cent); Kilkenny (27.1 per cent) and Kildare (26.7 per cent). At the other end of the distribution, the counties with the lowest percentage of their population in this social category included Leitrim (17.8 per cent); Donegal (18.6 per cent); Monaghan (19.0 per cent) and Cavan (19.2 per cent).

Table 3.5 also shows the percentage of each county's population in the Unskilled Manual class. From column 3 of the table we can see that Wexford has the highest rate (14.7 per cent), followed by Donegal (14.3 per cent), Waterford (13.8 per cent); and Carlow (13.7 per cent). At the other end of the distribution we have Dublin County (6.7 per cent); Roscommon (7.5 per cent) and Sligo (8.4 per cent). A comparison of columns 2 and 3 in the table indicates that there is clearly an inverse relationship between the rate of persons in the Professional and Unskilled Manual classes at the county level. At the county level the correlation coefficient between the percentage in the combined professional classes on the one hand and the unskilled manual class on the other is -0.2314. At the Rural District level it is -0.2244 and at the DED level it is -0.3761.

Table 3.5: Rates and Incidence of Higher and Lower Professional and Unskilled Manual Social Classes by County, 1991

Region	% Higher & Lower Professional (Rank)		% Unskilled Manual (Rank)		Distribution of		Total Population
					Higher & Lower Professional	Unskilled Manual	
East Region							
Dublin Co. Borough	21.7	(20)	11.1	(12)	11.7	14.5	13.6
Dublin County	34.4	(1)	6.7	(28)	21.2	9.9	15.5
Kildare	26.7	(4)	10.2	(19)	3.7	3.4	3.5
Meath	26.3	(5)	12.0	(7)	3.1	3.4	3.0
Wicklow	29.4	(2)	11.1	(13)	3.2	2.9	2.8
South-West Region							
Cork	26.0	(6)	10.3	(18)	12.0	11.5	11.6
Kerry	22.1	(18)	11.8	(8)	3.0	3.9	3.5
South East Region							
Carlow	22.9	(17)	13.7	(4)	1.1	1.5	1.2
Kilkenny	27.1	(3)	10.7	(15)	2.2	2.1	2.1
Wexford	23.4	(14)	14.7	(1)	2.7	4.1	2.9
South Tipperary	23.3	(15)	13.3	(5)	2.0	2.7	2.1
Waterford	24.0	(12)	13.8	(3)	2.5	3.5	2.6

North-East Region					
Louth	21.8	(19)	11.4	(10)	2.2	2.8	2.6
Cavan	19.2	(25)	10.5	(16)	1.1	1.5	1.5
Monaghan	19.0	(26)	9.4	(23)	1.1	1.3	1.5
Mid-West Region					
Clare	24.9	(8)	10.2	(20)	2.6	2.5	2.6
Limerick	24.3	(10)	11.0	(14)	4.4	4.8	4.6
North Tipperary	25.9	(7)	11.5	(9)	1.7	1.8	1.6
Midlands Region					
Laois	22.9	(16)	11.1	(11)	1.3	1.6	1.5
Longford	20.2	(23)	10.3	(17)	0.7	0.9	0.9
Offaly	20.9	(22)	13.0	(6)	1.4	2.1	1.7
Westmeath	24.7	(9)	10.0	(21)	1.7	1.7	1.8
Roscommon	21.0	(21)	7.5	(28)	1.2	1.1	1.5
West Region					
Galway	23.9	(13)	9.0	(25)	4.9	4.4	5.1
Mayo	19.4	(24)	9.9	(22)	2.4	3.0	3.1
North-West Region					
Leitrim	17.8	(28)	9.3	(24)	0.5	0.6	0.7
Sligo	24.1	(11)	8.4	(26)	1.5	1.3	1.5
Donegal	18.6	(27)	14.3	(2)	2.7	5.0	3.6
STATE	25.2	(-)	10.4	(-)	100.00	100.00	100.00

The distribution of these *rates* at the Rural District level for both classes are shown in Figures 3.1 and 3.2. Figure 3.1 shows that the percentage of persons in the combined professional categories are more densely located in the DEDs in the south and east of the country, to the south-east of a line drawn from Co. Louth to Galway, with outliers in Sligo and Castlebar.

Figure 3.1: Per Cent Higher and Lower Professional, 1991

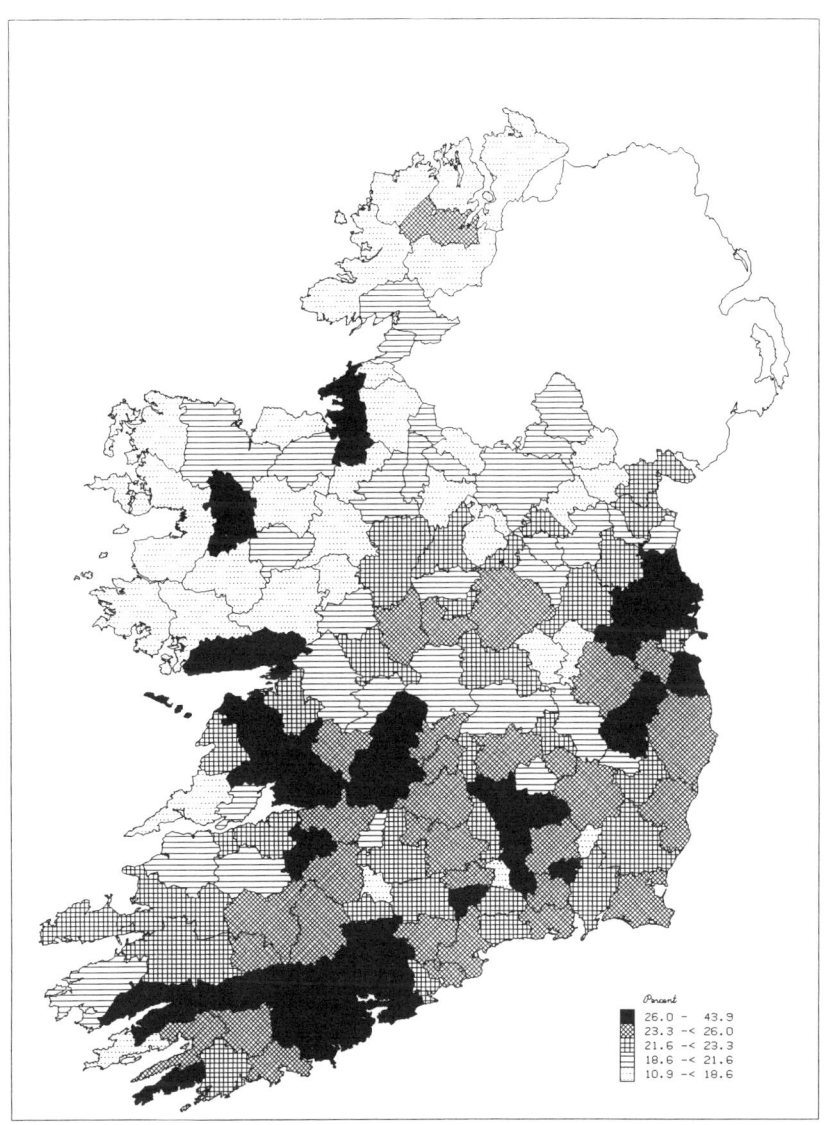

Figure 3.2: Per Cent of Unskilled Manual, 1991

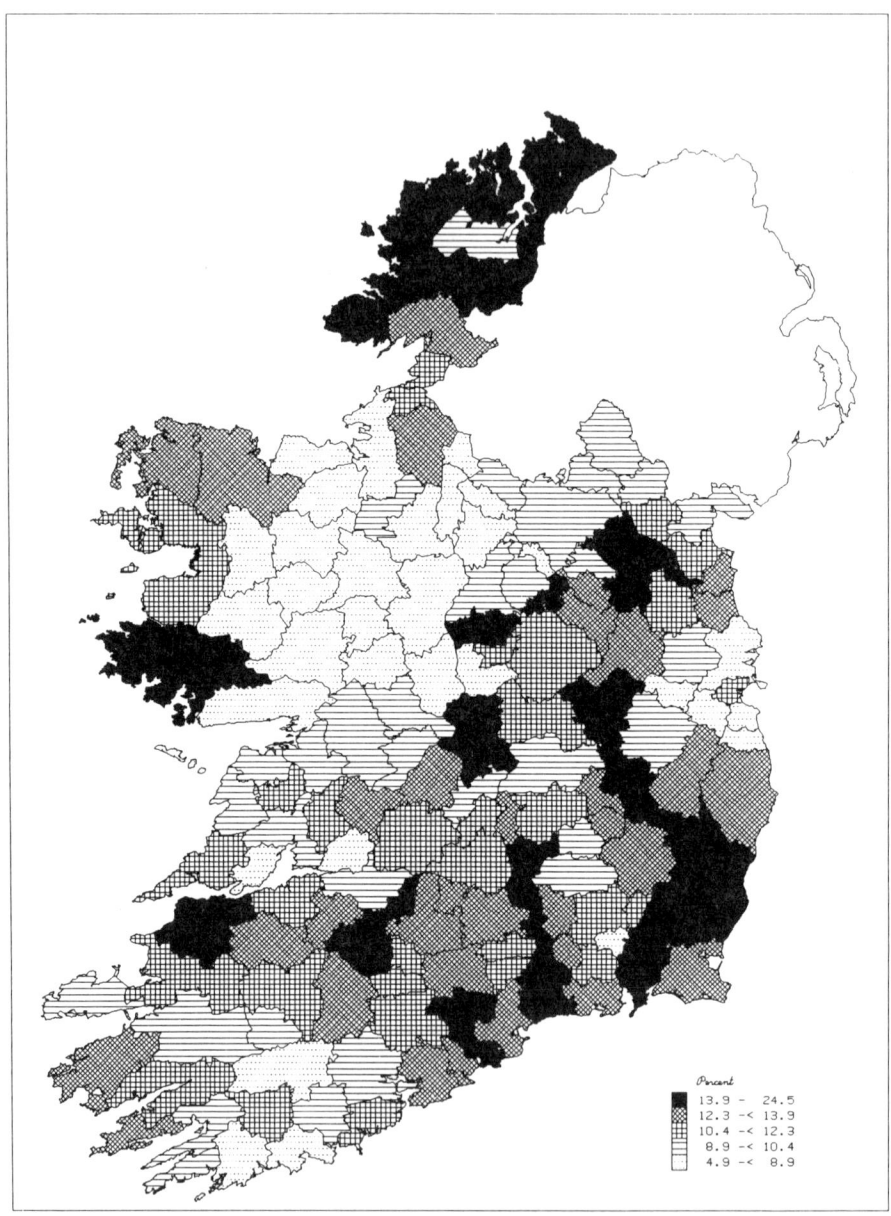

Figure 3.2 shows that there is a slightly greater degree of dispersal throughout the county of RDs which have a high rate of

persons in the Unskilled Manual category. Perhaps the most striking aspect of this map is the relative absence of RDs with a high percentage of their population in the Unskilled Manual group in parts of counties Roscommon, Longford, Mayo, Sligo, Leitrim and Cavan. This trend is largely attributable to the importance of farmers, and particularly small farmers in the occupational structure of the areas in question. The Census classification assigns farmers to social class according to size of farm. Small farmers (less than 30 acres) are assigned to the Semi-Skilled Manual category. No farmers (of any size) are assigned to the Unskilled Manual category. The highest incidence of small farmers (less than 30 acres) is in the Roscommon, Leitrim, Cavan areas. This mans that, by definition, such people do not fall into the Unskilled Manual class category. This results in the relative absence of RDs with a high percentage of their population falling into the relevant class category in the counties in question.

To provide some insight into the degree of concentration or otherwise of both classes at a sub-county level we can look at the incidence or distribution of both groups according to their relevant decile distribution of rates. For example, if we first consider the combined professional group we can examine how people in this class category are distributed across the country according to the deciles formed by DED rate of the professional group. In other words, the rate of occurrence of persons from the combined professional classes was calculated for each of the 3,440 DEDs in the country. Deciles of DEDs were then formed on the basis of those *rates*. The distribution of the absolute number of persons from the professional classes in each of those deciles was then calculated. The distribution of all persons was also derived according to the areas formed by the decile *rates* of the Higher and Lower Professional classes. This can be used as a reference benchmark against which to measure the degree of concentration of the professional categories. Comparable distributions were extracted in respect of the Unskilled Manual class. These distributions are presented in Table 3.6. The reader should note that all considerations of spatial contiguity have been ignored in constructing this table.

Table 3.6: National Distribution of Persons in Higher and Lower Professional and Unskilled Manual Classes Classified by Relevant Decile Rate at District Electoral Division Level in 1991

	Higher & Lower Prof.	All Persons	Unskilled Manual	All Persons
	%		%	
Decile 1 (low rate)	3.0	10.0	3.7	13.6
Decile 2	3.7	7.2	5.7	10.8
Decile 3	4.9	7.7	7.0	10.5
Decile 4	7.0	9.5	7.0	8.7
Decile 5	6.7	8.1	9.4	10.0
Decile 6	9.5	10.4	9.8	9.3
Decile 7	9.7	9.7	11.3	9.4
Decile 8	10.4	9.4	14.8	10.9
Decile 9	14.0	10.9	13.2	8.1
Decile 10 (high rate)	31.1	17.2	18.0	8.6
Total	100.0	100.0	100.0	100.0

From the first column of Table 3.6 we can see that 31 per cent of those in the professional categories were located in the 10 per cent of DEDs across the country with the highest *rate* of professional persons. The same areas contained a total of 17 per cent of the population, indicating an over-representation in the area formed by the top 10 per cent of *rates* in the order of 81 per cent. The top three deciles contain 55 per cent of persons from the two professional class categories compared with 37 per cent of all persons (an over-representation of 48 per cent). This over-representation of persons in the classes in question at the upper end of the rates distribution is compensated for at the lower end. For example, only 3 per cent of those in the professional group are found in the 10 per cent of DEDs formed by the distribution of the relevant *rate*. This compares with the 10 per cent of the total population who are located in this lowest decile.

A comparable distribution of persons from the Unskilled Manual class category is apparent from columns 3 and 4 of Table 3.6. For example, 18 per cent of persons in this grouping are located in the top decile compared with 9 per cent of the population as a whole. This represents an over-concentration in the order of 109 per cent.

Class Distribution in Dublin

We now turn to consider spatial variations in these class distributions in Dublin. In the city and county as a whole the overall percentage of persons classified as being in the Higher and Lower Professional categories was 28.5. A total of 8.7 was classified as Unskilled Manual. Ward levels varied substantially for both class categories. Figure 3.3 shows the distribution of rates for the higher and lower professional categories. Rates varied across the wards from a low of 3.1 per cent in Priorswood B in Darndale to the Foxrock-Torquay ward which had a figure of 67.2 per cent of its population in the professional class categories. In general, one can see that there is a relative concentration of wards with high percentages of their population in the professional groups running through the south-east of the County Borough area to the Pembroke wards in the Ballsbridge, Sandymount and Donnybrook areas as well as parts of Rathmines, Rathgar, Terenure and Rathfarnham. This region of high rate wards continues into the county area through parts of Templeogue, Churchtown, Dundrum and into south-east County Dublin in Dun Laoghaire-Rathdown. To the north of the county we can see outliers with a high percentage of population classified as professional in the Castleknock-Park and Knockmaroon wards as well as Howth and Malahide.

Figure 3.4 shows the percentage of persons in each ward assigned to the Unskilled Manual category. To some extent the distribution in this map is a mirror image of that in Figure 3.3. The map shows that rates vary from a low of 0.75 per cent in the Terenure C ward (with a total population of 1,859) to a high of 31.94 in the Merchants Quay A ward (with a total population of 1,124). A central corridor of high rate areas is apparent, running through the inner city County Borough area with an outer ring in parts of Finglas, Ballymun and Darndale. In the north County area (Fingal) we can see a tract of high rate wards running north-south through parts of Swords, through Hollywood, Naul and Balscadden to the north. In the south County area, parts of Tallaght, Firhouse and Bohernabreena stand out as having extremely high rates of their population in the Unskilled Manual class.

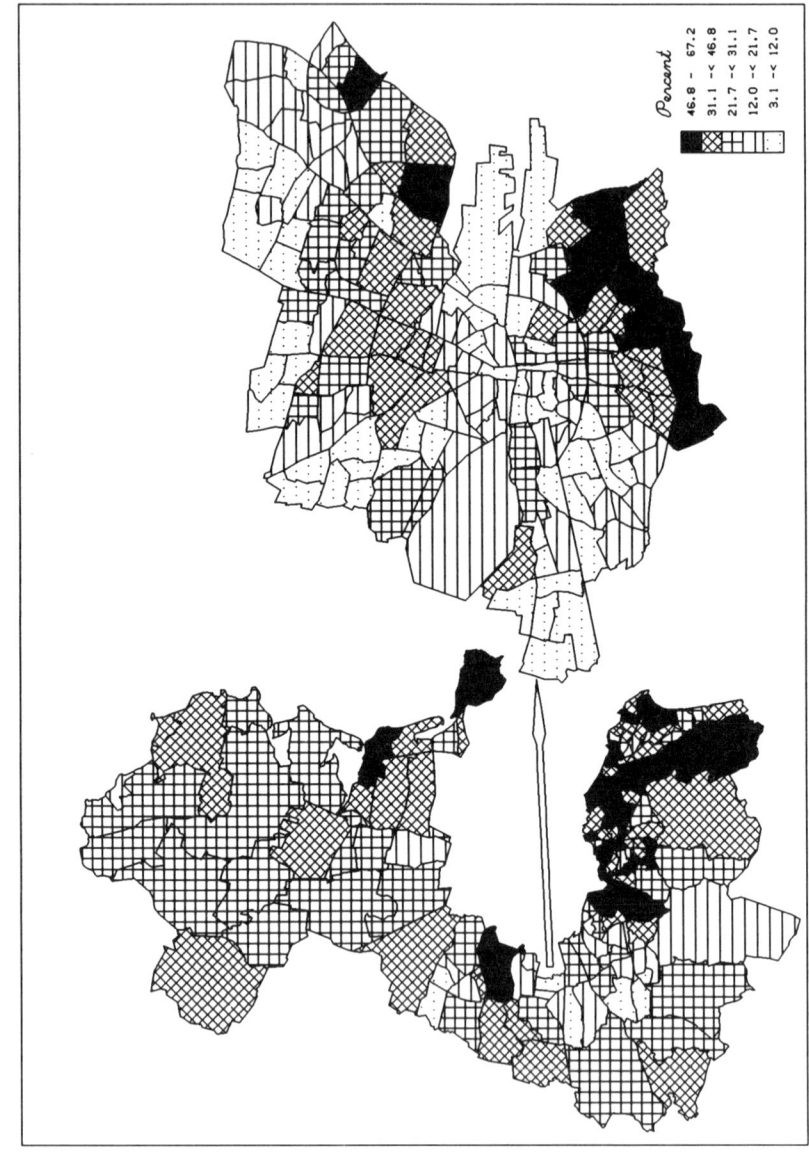

Figure 3.3: Dublin County & County Borough: Percentage in Higher and Lower Professional, 1991

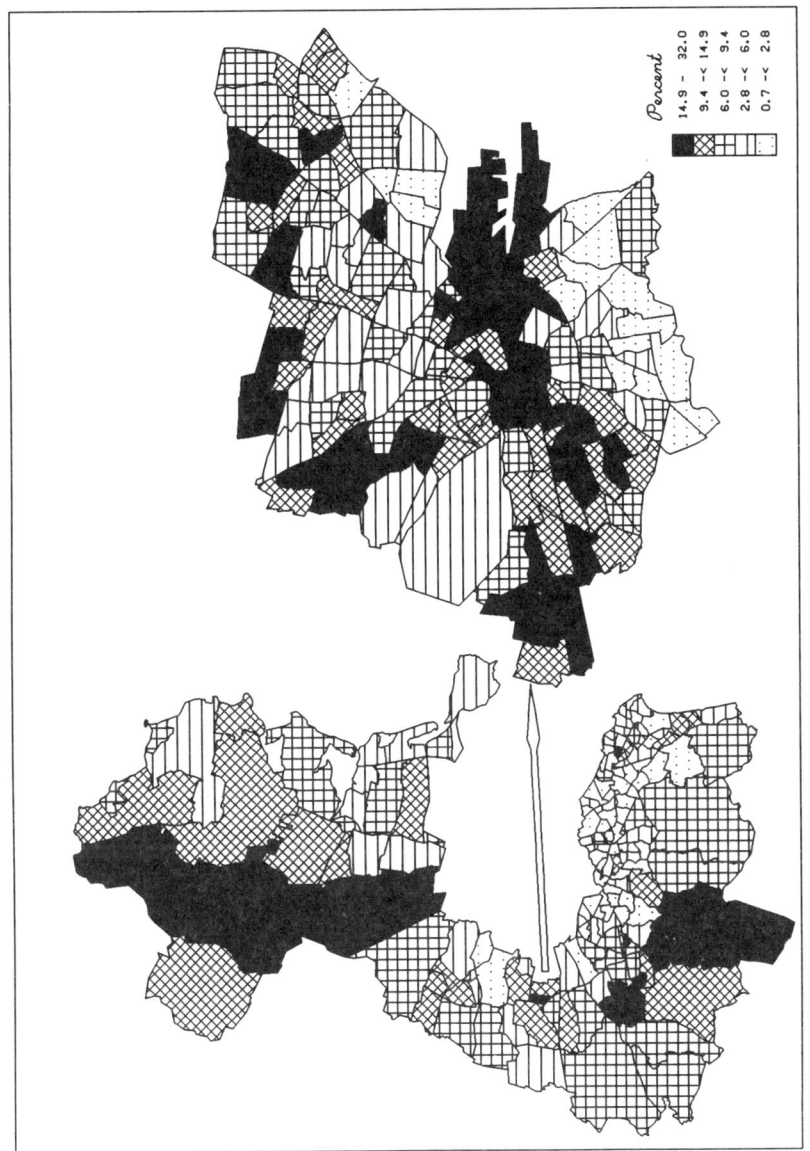

Figure 3.4: Dublin County & County Borough: Percentage in Unskilled Manual Class, 1991

When we compare the distributions in Figure 3.3 and 3.4, we can see that, as noted, each is largely a reflection of the other. The strength of this negative relationship between the percentage of a ward's population which is classified as being in the professional class and the percentage classified as unskilled manual is summarised in a correlation coefficient of −0.8285 (significant at the 99 per cent level) between the two variables. By any standard this represents a strong negative relationship between the distribution of the two class categories in question.

Turning from rate to incidence we can consider the distribution of both professional and unskilled manual categories according to the areas formed by the respective decile rate distributions of the two variables in question. This will indicate the degree to which persons in each class category are concentrated (or otherwise) within the areas of highest/lowest rates. The information is presented in Table 3.7. The first two columns concentrate on the persons in the higher and lower professional class categories. It shows that approximately 19 per cent of those in this group are located in the areas formed by the 10 per cent of wards with the highest *rate*. This same group of wards contain 9 per cent of all persons, suggesting an over-concentration of the professional group in this area in the order of 104 per cent. The table shows that the 30 per cent of wards with the highest *rates* of professional persons contain approximately 53 per cent of all persons classified as being in the higher or lower professional groups. The same 30 per cent of wards contain exactly 30 per cent of all persons in the population — an over-representation in the order of 77 per cent.

The final two columns in Table 3.7 deal with the distribution of the Unskilled Manual category, classified by decile rate. From this we can see that the 10 per cent of wards with the highest *rates* contained 22 per cent of the Unskilled Manual group. These same areas contain 9 per cent of all persons as a whole. This represents an over-concentration of the unskilled manual group of approximately 142 per cent in the areas formed by the top decile of rates. Wards in the top three deciles contain 54 per cent of the unskilled class, compared with 27.8 per cent of all persons — an over-representation of 94 per cent.

Table 3.7: Distribution within Dublin City and County of Persons in Higher and Lower Professional and Unskilled Manual Classes Classified by Relevant Decile Rate at the Ward/DED Level in 1991

	Higher & Lower Prof.	All Persons	Unskilled Manual	All Persons
	%		%	
Decile 1 (low rate)	2.0	10.0	1.4	8.6
Decile 2	3.4	10.3	2.6	10.1
Decile 3	5.2	10.3	4.7	11.4
Decile 4	6.9	10.0	6.3	10.5
Decile 5	8.3	10.1	9.4	12.0
Decile 6	9.6	9.6	9.4	9.6
Decile 7	11.9	9.8	12.2	10.0
Decile 8	15.8	10.5	16.3	10.3
Decile 9	18.4	10.4	16.2	8.4
Decile 10 (high rate)	18.6	9.1	22.0	9.1
Total	100.0	100.0	100.0	100.0

Unemployment

National Trends in Unemployment

A second important proxy of disadvantage and deprivation is unemployment. Table 3.8 presents details on regional variations in employment rates at the county level. From this we can see that the national unemployment rate from the 1991 Census of Population was 16.9 per cent. In terms of individual counties we can see that Donegal had the highest rate (25.4 per cent), followed by Louth (22.2 per cent) and Dublin Co. Borough (21.7 per cent). Counties with the lowest level of unemployment included Roscommon, Cavan, Clare and Leitrim, all with rates in the range of 10.2 to 13.7 per cent.[9]

Table 3.8: Rates and Incidence of Unemployment by Planning Regions and Counties

Region	%	(Rank)	Unemployed	Aged 15+
East Region				
Dublin Co. Borough	21.7	(3)	19.9	14.8
Dublin County	14.1	(24)	13.4	15.1
Kildare	14.4	(21)	2.9	3.3
Meath	15.3	(14)	2.6	2.9
Wicklow	17.7	(9)	2.8	2.7
South-West Region				
Cork	15.2	(16)	10.2	11.7
Kerry	16.7	(11)	3.2	3.5
South East Region				
Carlow	18.4	(5)	1.2	1.1
Kilkenny	15.1	(18)	1.8	2.0
Wexford	19.2	(4)	3.1	2.8
South Tipperary	18.1	(7)	2.2	2.1
Waterford	18.2	(6)	2.8	2.6
North-East Region				
Louth	22.2	(2)	3.3	2.5
Cavan	12.6	(27)	1.1	1.5
Monaghan	14.2	(23)	1.2	1.4
Mid-West Region				
Clare	13.3	(26)	2.0	2.6
Limerick	17.8	(8)	4.8	4.6
North Tipperary	15.1	(17)	1.4	1.6
Midlands Region				
Laois	16.2	(12)	1.3	1.4
Longford	14.9	(19)	0.7	0.8
Offaly	17.1	(10)	1.6	1.6
Westmeath	14.3	(22)	1.4	1.7
Roscommon	10.2	(28)	0.9	1.5
West Region				
Galway	15.3	(15)	4.5	5.1
Mayo	16.0	(13)	2.7	3.1

North-West Region				
Leitrim	13.7	(25)	0.5	0.7
Sligo	14.8	(20)	1.3	1.6
Donegal	25.4	(1)	5.1	3.5
State	16.9	-	-	-

Rural district unemployment rates are shown in Figure 3.5.

Figure 3.5: Unemployment Rates, 1991

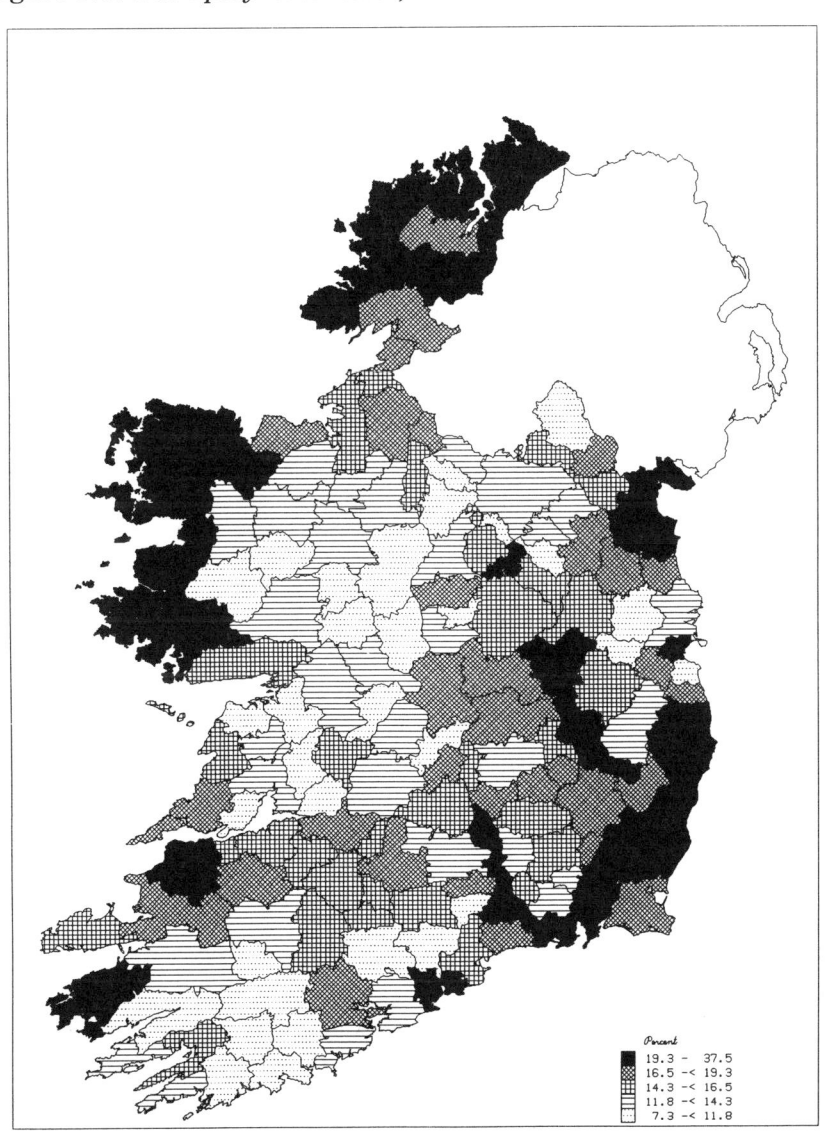

One can see that, in general, RDs with the highest levels of unemployment were in parts of Co. Donegal, north and west Mayo into parts of west Galway, north and south-west Kerry and a long tract through parts of counties Wicklow, Wexford, Waterford, Kilkenny, Kildare and Offaly.

Whilst the identification of these disadvantaged areas (using the unemployment rate criterion) is a valuable exercise, it is once again essential to draw the distinction between risk and incidence. Risk figures in isolation could give a distorted picture of what is happening on the ground if, for example, a high risk level is derived from a small labour force base. In addition to identifying areas of "high risk", therefore, one must also consider the percentage of the poor or disadvantaged who are found in these "high risk" areas. The last two columns of Table 3.8 change the focus from risk to incidence at the county level. Column 3 provides a breakdown of the unemployed across the counties while column 4 provides a comparable breakdown for the distribution of persons aged 15 years or more. This latter set of figures acts as a reference group against which to compare the distribution of the unemployed. The most striking aspect of the table is the degree to which the distribution of the unemployed reflects the overall distribution of the reference population (those aged 15 years and over). The ratio of column 4 ranges from a high of 1.46 for Donegal to a low of 0.6 for Roscommon. This implies that Donegal has 46 per cent more unemployed than it would have if the unemployed were distributed on an equitable pro rata basis in line with the general population aged 15 years and over. Other counties with a ratio in excess of unity include Louth (1.32), Dublin Co. Borough (1.34) and Wexford (1.11). It should be noted that the disproportionate over-representation in Dublin County Borough is compensated for by an under-representation in the Dublin County area. The ratio of the unemployed to the population aged 15 years and over in Dublin County is 0.89, suggesting an under-representation of 11 per cent. When the Dublin County and County Borough regions are treated as single unit we can see that County Dublin contains exactly 33.3 per cent of the unemployed compared with 29.9 per cent of the population aged 15 years and over. This suggests, therefore, that the unemployed are over-represented in Dublin as a whole by a factor of approximately 11 per cent.

Spatial Aspects of Poverty and Deprivation

The county-level data clearly mask a substantial degree of internal variation in unemployment risk and incidence figures at a lower level of spatial disaggregation. To get some insight into the degree of concentration of the unemployed within the areas of highest risk we can look at the incidence of unemployment across the decile distribution of unemployment risk. Table 3.9 presents the relevant information. This table contains details on the national distribution of the unemployed classified by decile unemployment rate at the District Electoral Division level in 1991. There is a total of approximately 3,440 DEDs in the country (including the wards within the five county boroughs). In Table 3.9 we have grouped the DEDs into deciles according to their unemployment rate. We then consider how many of the unemployed (and the reference category of those aged 15 years and more) would be contained within the areas defined in terms of these deciles, i.e., defined in terms of degree of risk of unemployment.

Table 3.9: National Distribution of Unemployed Classified by Decile Unemployment Rate at District Electoral Division Level in 1991

Unemployment Rate Decile	Percentage of the Unemployed	Percentage of Total Population Aged 15+
	%	
Decile 1 (low rate)	2.3	6.6
Decile 2	3.8	8.0
Decile 3	5.1	8.9
Decile 4	6.3	9.4
Decile 5	6.6	9.0
Decile 6	7.5	9.0
Decile 7	9.9	10.5
Decile 8	12.3	11.3
Decile 9	16.1	12.0
Decile 10 (high rate)	30.1	15.3
Total	100.0	100.0

All considerations of spatial contiguity are ignored in the table. From Table 3.9 we can see that the DEDs in the top decile of unemployment rates contained just over 30 per cent of the unemployed, compared with 15.3 per cent of the population aged 15 years or more. This represents an over-concentration of almost 100 per cent as compared with the situation which would pertain were the unemployed to be distributed across the DEDs in line with the distribution of the population aged 15 years or more. The table shows that the 30 per cent of DEDs with the highest *rates* of unemployment contain 58.5 per cent of the unemployed. They also contain 38.6 per cent of the population aged 15 years and over. This implies an over-representation of the unemployed in the high risk areas in question of 51.5 per cent. Although this is clearly a very substantial level of over-representation the reader should note that if the objective of policy is to reach a majority of the disadvantaged group (as measured in this case in isolation by unemployment) one would achieve this objective only by encompassing an area which also contained 39 per cent of the adult population.

Trends in Unemployment in Dublin
What of the situation regarding unemployment in Dublin? The 1991 Census returned a rate of 17.8 per cent for the city and county as a whole. Ward-level rates display a very substantial degree of variation, however, ranging from a maximum of 60.5 per cent in the Mountjoy A ward (representing a total of 672 persons unemployment from a total labour force of 1,100 persons) to a minimum of 3.9 per cent in the Castleknock-Park ward (representing a total of 71 unemployed from a labour force of 1,838). The degree in variation in rates can be seen from Table 3.10. From this we can see that 61 per cent are below the aggregate rate for the city and county as a whole while 10 per cent of the wards are more than twice the aggregate figure. Wards which experience an extremely high rate of unemployment include Priorswood C (59.8 per cent); North Dock C (58.1 per cent); Blanchardstown-Mulhuddart (55.6 per cent); Priorswood B (55.5 per cent); and Mansion House B (54 per cent). At the other end of the distribution, wards with a particularly low rate include Clonskeagh-Belfield (4.3 per cent); Firhouse-Ballycullen (4.3 per cent); Rathfarnham village (4.8 per cent); Stillorgan-Mount Merrion (5.0 per cent); and Stillorgan-Deerpark (5.1 per cent).

Table 3.10: Distribution of Unemployment Rates in the 322 Wards of Dublin City and County in 1991

Unemployment Rate	Per Cent of Ward
LE 9%	22.0
> 9-18%	39.1
> 18-27%	15.8
> 27-36%	13.0
> 36%	9.9
Total	100.0

Figure 3.6 shows the unemployment rate at the ward level for Dublin city and county. From this one can see that a central corridor of wards which experience the highest risk of unemployment stretches from east to west through the County Borough in the North Inner City area from North Dock A, through parts of Ballybough and along the quays. On the southside it extends from the Mansion House A ward, along the quays through parts of Kilmainham, Crumlin and Drimnagh, extending into Inchicore, Ballyfermot and Cherry Orchard. There is also an outer ring of high risk areas in the north of the County Borough, containing parts of Finglas, Ballymun and Darndale (especially in the Priorswood wards). To the west of the city boundary the east-west corridor extends into the county area into parts of Clondalkin and Tallaght (particularly the Fettercairn and Jobstown wards). To the northwest we can see that parts of Blanchardstown — especially the Coolmine, Corduff, Mulhuddart and Blakestown areas stand out with exceptionally high rates of unemployment.

Shifting the focus again from rate to incidence, we turn finally, in Table 3.11, to look at the distribution of unemployed persons in Dublin city and county classified by decile unemployment rate at the ward level. This shows, for example, that just over 20 per cent of unemployed persons live in the wards which constitute the area formed by this top decile of unemployment rates. The table shows that this group of wards contains only 8 per cent of the total population aged 15 years or more. This indicates an overrepresentation of unemployed persons in these high risk areas of the order of 154 per cent. If one's policy objective is to target over

50 per cent of the unemployed in Dublin one would have to include the area formed by the top three deciles of unemployment rates. These 30 per cent of wards contain 52 per cent of the unemployed population, compared with 28 per cent of all persons aged 15 years and over. This represents an over-concentration of the order of 87 per cent in the wards in question.[10]

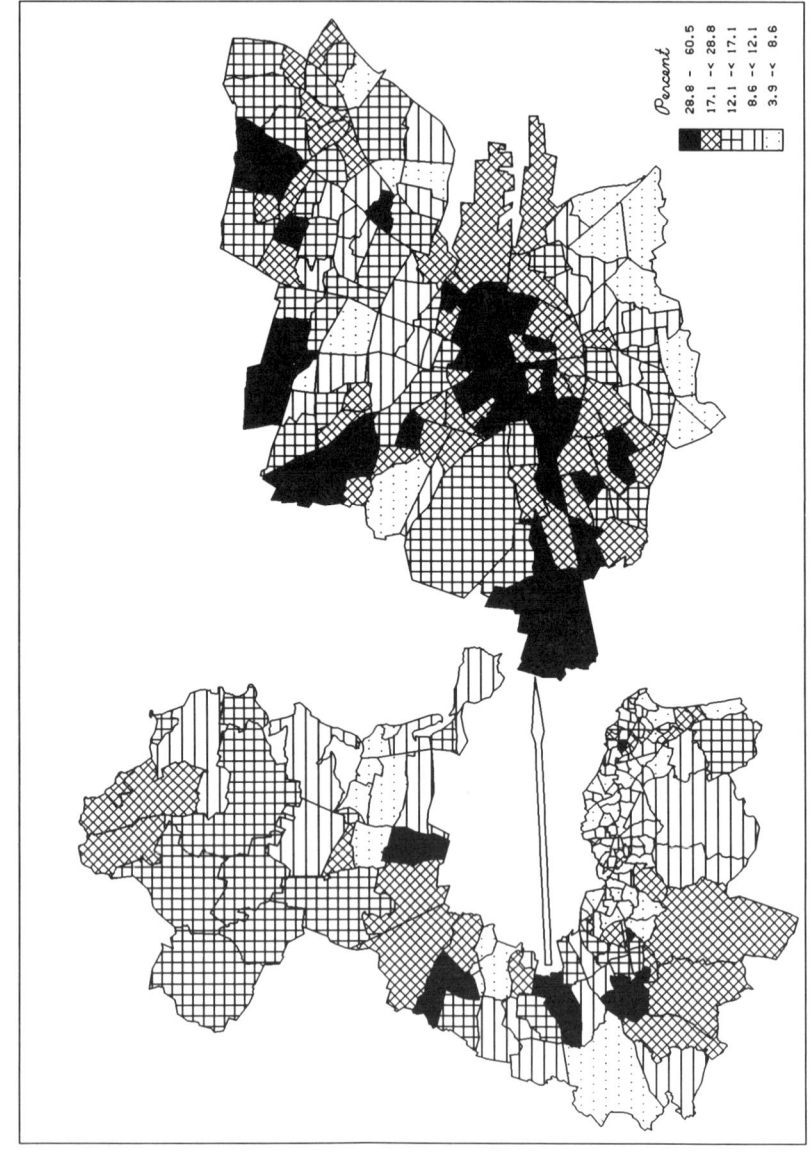

Figure 3.6: Dublin County & County Borough: Unemployment Rates, 1991

Table 3.11: Distribution of the Unemployed in Dublin City and County Classified by Decile Unemployment Rate at the Ward/DED Level in 1991

Unemployment Rate Decile	Percentage of the Unemployed	Percentage of Total Population Aged 15+
Decile 1 (low rate)	3.5	10.3
Decile 2	4.6	10.7
Decile 3	5.7	10.5
Decile 4	7.4	11.5
Decile 5	8.1	10.6
Decile 6	8.9	10.2
Decile 7	9.5	8.2
Decile 8	14.8	10.2
Decile 9	17.4	9.9
Decile 10 (high rate)	20.1	7.9
Total	100.0	100.0

N.B. This table relates to the 322 wards in Dublin city and county.

CONCLUSIONS

The purpose of this chapter was to adopt a spatial perspective on poverty and deprivation in Ireland. This is an especially relevant topic for policymakers at this time, given the increased role of area-based responses to poverty both at a national level within Ireland and also at a wider level, e.g., within the EU under its various poverty programmes. There is currently a host of local development groups, both voluntary and State-sponsored, which are involved in local development initiatives of one sort or another. At the core of this area-based response are the 38 partnerships, which were initially set up in 1991 as the 12 original PESP areas. In addition, we have other development organisations working at the local level. These include the County Enterprise Boards, Leader Projects and, most recently, the County Strategy Groups which were set up as co-ordinating bodies under the Operational Programme for Local Urban and Rural Development. Work such as that presented in this chapter is clearly central to the job facing these local development initiatives. In particular, the results presented can inform the debate on the degree of concentration or otherwise of poverty and deprivation as well as throwing light on the relevance or otherwise or urban underclass arguments in an Irish context.

Because much of this research is based on cross-sectional survey data, one is limited in the degree to which one can disaggregate the results. Nonetheless, quite a lot can be learned by examining poverty risk and incidence levels according to a classification based on area-type as well as one based on the nature of household tenure.

We firstly considered poverty risk figures according to the area-type classification. This showed that lowest risk was experienced by households in open country (8.3 per cent) followed by Dublin City and County (14.0 per cent). Highest risk was experienced by households in the four county boroughs as well as those in small towns or villages (19.4 per cent and 20.6 per cent, respectively). These figures alone suggest that poverty is not the sole preserve of large urban areas but is instead widely dispersed throughout the country.

Using the combined 60 per cent relative income plus basic resource deprivation line, we saw that Dublin City and County contained a total of 30.1 per cent of households in poverty. This same area contains a total of 29.5 per cent of all households in the State. Further, the other four county boroughs contained 12.4 per cent of households in poverty, compared with 8.7 per cent of all households. Although these figures clearly suggest a degree of over-concentration in the large urban areas outside Dublin, they do not indicate that poverty has become exclusively ghettoised in the areas in question. Indeed, we saw that both medium-sized towns as well as small towns/villages each had an over-concentration of households below the poverty line.

The analysis of area-type was complemented with one based on the nature of tenure. We saw that risk figures for the local authority rental sector were substantially higher than for other tenure types. A total of 47 per cent of households in that sector were in poverty, compared with 13.7 per cent of all households in the State. The local authority tenant-purchase sector had the second highest risk levels at just over 23 per cent. These figures compare with 7.2 per cent for the private sector with an outstanding mortgage. In terms of incidence we found that 51 per cent of poor households were in local authority housing as compared with 19.7 per cent of all households in the state.

When we combined the area-type and tenure variables, we found that poverty risk among local authority households was substantially higher than among households in the private housing sector *across all area types*. The disparity in risk levels between public and private housing was lowest in areas of open country, where the risk figure for local authority is 14 per cent, compared with a figure of 8 per cent among other households in these areas. This represents a ratio between local authority/private sector households of 1.75:1. The ratio of risk in other areas is much higher. In the four county boroughs, for example, it is 6.0:1 and reaches a maximum of 8.0:1 in Dublin City and County.

In terms of poverty incidence cross-classified by tenure and area-type, we found that just over 30 per cent of all poor households are in the local authority sector in Dublin and the other four county boroughs. This compares with a figure of 9.7 per cent for all households of this type (i.e., local authority in Dublin and the other four county boroughs) in the population as a whole. This clearly represents a substantial concentration of poverty in local authority housing in the capital and other county boroughs. One must remember, however, that it also says that 70 per cent of households below the poverty line in 1994 were located in other area-type/tenure categories throughout the country.

To provide some more spatially disaggregated insights into the risk and incidence of poverty we presented some limited analysis of social class and unemployment from the SAPS data. When discussing the latter we saw that the counties with the highest rates include Donegal (25.4 per cent), Louth (22.2 per cent) and Dublin County Borough (21.7 per cent). At the other end of the distribution we identified counties Roscommon, Cavan, Clare and Leitrim with unemployment rates in the range 10-14 per cent. The importance of farming, particularly small farming, in the occupational structures of the areas in question must be taken into account in interpreting these low unemployment rates.

When we considered the incidence of unemployment at the county level the most striking feature was the degree to which the distribution of the unemployed reflected that of the reference population i.e. those aged 15 years and over. There were clearly individual counties which experienced an over- or under-representation in terms of the distribution of unemployed. For example, we saw that Donegal had an over-representation of the order of 46 per cent; Louth an over-representation of 32 per cent; Dublin County Borough an over-representation of 34 per cent. We noted, however, that when Dublin County and County Borough were treated as a single unit the county as a whole contained 33.3 per cent of the unemployed, compared with 29.9 per cent of the population aged 15 years and over. This highlights the problem of modifiable areal units in spatial analysis of this nature.

We also examined the *distribution* of the unemployed classified by decile rate of unemployment at the DED level. DEDs in the area formed by the top decile of rates contained 30 per cent of the unemployed, compared with 15.3 per cent of the population aged 15 years or over. This represents an over-concentration of almost 100 per cent as compared with the situation which would pertain were the unemployed to be distributed across the DEDs in line with the distribution of the adult population. The top three deciles of DED unemployment rates contained 51 per cent of the unemployed but also 39 per cent of the population aged 15 years or more. This implies that if our policy objective was to target a majority of the unemployed one could do so only by encompassing 39 per cent of the adult population. If this were to be used as a criterion for policy implementation, one issue which would have to be addressed is the fate of the other 50 per cent of the unemployed who live outside the area formed by the top three deciles of unemployment rates.

When we focused on unemployment in Dublin City and County in isolation we found that the aggregate rate was 17.8 per cent. This ranged from a minimum of 3.9 per cent for the Castleknock-Park ward to a maximum of 60.5 per cent for Mountjoy A. We saw that we could identify a central corridor of high risk running through the inner city wards of the County Borough with concentrations of high risk areas also evident in parts of Finglas, Ballymun and Darndale as well as parts of Crumlin; Inchicore and Ballyfermot. Other areas of high unemployment risk in Dublin County included parts of Clondalkin and Tallaght as well as parts of Blanchardstown, Coolmine, Corduff, Mulhuddart and Blakestown.

Finally, turning to incidence of unemployment in Dublin, we saw that 20 per cent of the unemployed lived in the top decile of rates, compared with only 8 per cent of the adult population — representing an over-representation in these high risk areas in the order of 154 per cent. If one's policy object was to target a majority of the unemployed in Dublin city and county one would have to include the top 30 per cent of wards with the highest unemployment rates, containing 52 per cent of the unemployed and 28 per cent of all persons aged 15 years and over.

How does one interpret this fairly wide range of findings? The first thing which one can say is that all the results presented suggest that poverty, deprivation and disadvantage are spatially pervasive phenomena affecting almost every part of the country and every type of area within the country. The overriding impression is that poverty is not the preserve of any one area or type of area — be it urban or rural.

It is particularly important to stress that we are not saying that there are no concentrations of poverty. What we are saying is that, on the basis of the research so far, poverty is a spatially pervasive phenomenon which affects all parts of the country. Depending on the scale one uses, pockets or concentrations of disadvantage can clearly be identified. We noted unemployment rates of up to 60 per cent in certain wards of Dublin. We identified a central tract of high rate areas in the County Borough region as well as an outer ring in parts of Finglas, Ballymun and Darndale. Areas of Blanchardstown and Tallaght were also highlighted as experiencing a high risk of disadvantage as proxied by unemployment rate. The impact of such high rates on the lives of the people involved cannot be overemphasised. The identification of these high risk areas and related concentrations to some extent depends on the scale at which we are operating and so highlight modifiable areal unit problems which affect all spatial analysis of this nature.

Secondly, it is hoped that the findings presented help to clarify the need to distinguish between process and outcomes. A failure to make this distinction could well impact adversely on the efficacy and efficiency of policy. If the processes underlying poverty are essentially structural, then area-based strategies clearly should not be the only policy instrument used. If the processes are structural then reforms in education and training systems will be important to allow second chance education and vocational training. One should of course note that even if one finds that structural processes are largely responsible for determining poverty, this of itself should not be seen to imply that policymakers should ignore the spatial — especially if the structural processes have the effect of creating geographic concentrations of poverty and disadvantage. The important point in research of this nature is to consider whether or not one is looking at process or outcome. Failure to do so may ultimately lead one down the road of spatial determinism.

Thirdly, where do we go from here with this type of analysis? Much more work needs to be done in linking the cross-sectional survey data and the SAPs data. Trends from the latter could possibly be more clearly identified using data transformation techniques to account for the very large differences in size of wards/DEDs. A wider range of variables should be examined than was possible for this chapter and a factor analytic approach should be used to combine independent variables.

Finally, what of area-based strategies? The analysis presented in this chapter is not a criticism of such strategies *per se*. Until more integrated and broadly-based research has been undertaken, the potential impact of such policies remains to be assessed. What the present analysis strongly suggests, however, is that, in view of the spatial pervasiveness of the problem, the development areas will have to be extremely extensive and inclusive to encompass a substantial proportion of the "target" population. Structural issues have been identified as being of extreme importance in determining a household's probability of being in poverty. As such, the most appropriate message would seem to be a reminder that area-based strategies are only one component of overall policy in this area.

NOTES

1. This chapter draws heavily on Nolan, Whelan and Williams (1996) which was funded by the Combat Poverty Agency.

2. Other types of poverty line include the following: the absolute or budget standard lines which are based on the price of a "necessary" basket of goods for families of different sizes and composition; the social security line which is based on the minimum standard of living rates implicit in the Social Welfare code; consensual poverty lines based on the income level perceived by society as representing an acceptable or adequate standard of living.

3. For a full discussion of how these underlying dimensions of deprivation were constructed, see Whelan *et al.*, (1991) and Nolan and Whelan (1996), especially chapters 4-6.

4. This survey constituted the Irish component of the first wave (1994) of Eurostat's longitudinal survey of households — the Europanel survey.

5. The level of these figures will change depending on the poverty line chosen. For a full comparison of the effects of changing the location of the line and/or equivalence scales, see Callan *et al.* (1996). For the effects of such changes on the area-based rates and incidence levels, see Nolan, Whelan and Williams (1996).

6. Total sample size was 4048 households. The minimum number of unweighted households in any region was 300.

7. See, for example, Rottman *et al.* (1982), Roche (1984), Callan *et al.*, (1989), Whelan *et al.*, (1991), Callan and Nolan (1994), Callan *et al.* (1996).

8. The information presented in Tables 3.5 and 3.8 is based on the county level. Tables 3.6, 3.7 and 3.9 are based on data at the Ward/District Electoral Division level. These represent the most disaggregated spatial units for which Census data are available. The maps presented in this section have been drawn at the Rural District level. There is a total of 217 Rural and Urban Districts (RDs and UDs) in the country as a whole. For convenience and clarity of presentation we have incorporated the UDs into their relevant RD as appropriate. This leaves us with a total of 159 areal units in each of the maps presented.

9. As noted above in our discussion of social class, farming, and particularly small scale farming, plays a fairly important role in the occupational structure of many of the counties with a low rate of unemployment. Some of the farming activity in question (especially small scale farming) may, however, mask a significant degree of underemployment. No measure of the latter is available from the Census material.

10. The reader is reminded that, as was the case with our earlier discussion on the national distribution of unemployed persons by DEDs, no constraints of spatial contiguity have been imposed on the figures presented in Table 3.10.

REFERENCES

Callan, T., Nolan, B., Whelan, B.J. and Hannan, D.F. with Creighton, S. (1989) *Poverty, Income and Welfare in Ireland*, General Research Series, Paper No. 146, The Economic and Social Research Institute, Dublin.

Callan, T., Nolan, B., and Whelan, C.T. (1993) "Resources, deprivation and the measurement of poverty", *Journal of Social Policy*, 22(2), 141-172.

Callan, T. and Nolan, B. eds. (1994) *Poverty and Policy in Ireland*, Gill and Macmillan, Dublin.

Callan, T., Nolan, B., Whelan, B.J., Whelan, C.T. and Williams, J. (1996) *Poverty in the 1990s: Evidence from the 1994 Living in Ireland Survey*, Report to the Combat Poverty Agency and Department of Social Welfare.

Mack, J., and Lansley, S. (1985) *Poor Britain*. Allen and Unwin, London.

Mayer, S. and Jenks, C. (1988) "Poverty and the Distribution of Material Hardship", *Journal of Human Resources*, 24(1), 88-114.

Nolan, B., Whelan, C.T. and Williams, J. (1994) "Spatial Aspects of Poverty and Disadvantage", in Callan, T. and Nolan, B. eds. *Poverty and Policy in Ireland*, Gill and Macmillan, Dublin.

Nolan, B. and Whelan, C.T. (1996) *Resources Deprivation and Poverty*, Clarendon Press, Dublin.

Nolan, B., Whelan C.T. and Williams, J. (1996) *Spatial Aspects of Poverty and Disadvantage: A Comparison of 1987 and 1994*, Report to the Combat Poverty Agency, Dublin.

Roche, J. (1984) *Poverty and Income Maintenance Policies in Ireland*, Institute of Public Administration, Dublin.

Rottman, D.B., Hannan, D.F., Hardiman, N. and Wiley, M. (1982) *The Distribution of Income in the Republic of Ireland: A Study of Social Class and Family-Cycle Inequalities*, General Research Series, Paper No. 109, The Economic and Social Research Institute, Dublin.

Townsend, P. (1979) *Poverty in the United Kingdom*, Penguin, Harmondsworth.

Whelan, C.T. Hannan, D.F. and Creighton, S. (1991) *Unemployment, Poverty and Psychological Distress*, General Research Series, Paper No. 150, The Economic and Social Research Institute, Dublin.

Williams, J. (1993) *Spatial Variations in Deprivation Surrogates*, Report to the Combat Poverty Agency, Dublin.

4

METHODOLOGICAL ISSUES IN A CROSS-BORDER INVESTIGATION OF POVERTY IN IRELAND

*Sally Cook, Adrian Moore and Michael Poole,
School of Environmental Studies, University of Ulster, Coleraine*[1]

INTRODUCTION

The Need for Cross-Border Small Area Analysis

From the public housing estates of Belfast and Dublin to the barren hillsides of Leitrim and Donegal, we can all think of well-known areas on this island with a reputation for poverty. Both urban deprivation and its rural counterpart not only have a high profile in contemporary Ireland, but they also form a recurring motif in Irish history — for example, in relation to famine and migration.

It has always been easier, however, to refer glibly to a Ballymurphy or a Ballymun than to undertake a comprehensive objective analysis of the geography of poverty in either the Irish Republic or Northern Ireland. This, in turn, means that it has always been simpler to target resources at a handful of high-profile places or at a broad regional belt like the West than to make a carefully detailed, unbiased selection of particularly deprived locations.

Nevertheless, some notable attempts have been made both north and south of the Border to generate listings of geographical places rank-ordered according to the severity of their deprivation. This kind of research — some of it undertaken purely as an academic exercise, some done by academics for public sector sponsors, and some carried out entirely by government — has been going on for a quarter of a century. In a sense, our project is firmly in this tradition of applying deprivation indicators, but it is also a new departure by virtue of being a specifically cross-border investigation.

A European Policy Perspective

In an era when the political and economic system in Europe is undergoing rapid change, striving towards the creation of an integrated single European Market, policymaking (both economic and social) is experiencing a significant transformation. Recognition of the disadvantages of wide economic disparities between and within member states has led to a major area-based policy initiative designed to redress imbalances, but despite such moves, a significant minority continue to suffer social and economic disadvantage. In particular, it has been recognised that peripheral border regions suffer most from the problems of economic and social marginalisation. This goes against a key EU policy objective of promoting economic and social cohesion.

Within Ireland, the economic necessity of survival in changing market conditions, and the increased opportunities for substantial grant aided financial assistance (mostly from the EU), have driven the recent spate of initiatives for cross-border co-operation — loosely working, it could be said, toward the creation of a single island economy. These initiatives have not concentrated only on purely economic measures; already, at the sub-regional and local levels, policies and initiatives are being developed to address a number of dimensions of poverty such as health and unemployment. In the north-west border region, for example, the possibility of providing integrated health care services is being considered, while many local community-based cross-border job creation schemes, tackling problems of long term unemployment, have been funded by EU programmes.

Such developments form the backdrop of the current study. It is necessary to understand the multifaceted nature of the processes which lead to poverty, which may differ between north and south, before effective policies can be developed to combat the problems. A first step however is to identify the spatial pattern and extent of the problems, especially if we wish to develop carefully targeted area-based policies in an all Ireland context.

Problems of Data Availability and Suitability

The advantage of tackling the spatial analysis of poverty by focusing on small geographical areas is that it reduces the probability of clusters of poverty being masked by the use of a coarse spatial mesh. However, there is a major problem with the investigation of small areas. That is, that just as there is a hierarchy of area-size, so there is a closely correlated hierarchy of data availability. At the national level, there is an immense amount of data available; for administrative units like counties or District Council Areas, there is a fairly large quantity; but, for the micro-level of enumeration districts and District Electoral Divisions, there is a severe information shortage. To a large extent, this problem is created by (a) the origins of most published data as a by-product of an administrative process, and (b) the fact that sample survey work is useless for a complete coverage of small geographical areas. The first factor is significant because few administrative needs appear to require really micro-level data, while the second implies that only data sources with a 100 per cent coverage can generate information for small areas. When further factors like restrictions on public release are taken into account, this means that, in practice, there is little more than population census data available for micro-level analysis.

These comments apply to any investigation using data for small geographical areas, but, in addition, the analysis we require to undertake involves a second major problem with respect to information availability. This arises specifically because of the international comparative nature of the exercise. Our micro-level work, which is seen as making a local comparison to supplement the trend towards the more macro-level international comparative research so apparent in contemporary social science, shares with that macro-research the problem of obtaining truly comparable data. This is probably inevitable when there is so little international co-ordination of data-gathering exercises, but it is particularly disappointing when there is no language barrier between the jurisdictions involved in the comparison.

Taken together, these problems of limited data availability at the bottom of the spatial hierarchy and the rarity of internationally comparable information are so serious that the focus of this paper will be on these methodological issues rather than on empirical results. The purpose is to provide some general guidance derived from our own experiences of making international comparisons on the basis of small area data. In this way, it is hoped that the paper is relevant not only for the investigation specifically of poverty in Ireland, but also for the broader theme of census-based comparative research.

SMALL AREA POVERTY INDICATORS

There is considerable disagreement on the conceptualisation of poverty, let alone the even more difficult issue of operationalising the concept in empirical measurement (Callan and Nolan, 1994). Quite apart from the availability of alternative income-oriented concepts, which essentially express potential deprivation, there is the equally important issue of actual deprivation. The former is an indication of resource constraint, while deprivation is a consumption-oriented expression of people's exclusion from the level of living which is viewed as basic in that society. The measurement of both the resource constraint and the social exclusion element requires special sample survey research.

Not surprisingly, no source offers the ideal data at a small-area level for the whole of Ireland. Instead, we must operate within the confines of a far from ideal data source, i.e. the Census of Population; and as this data source contains no direct information on income or financial circumstances, it is necessary to rely on surrogate variables which are considered to indicate the presence or the likelihood of various forms of poverty or deprivation. Of course, the practice of using census data to categorise the proportion of a population experiencing deprivation is considerably less clear-cut than defining the risk of poverty according to income levels, even if the income levels employed may be considered somewhat arbitrary ones.

In discussing the methodological background to their British work in the field, Bradford *et al.* (1995) point out that there are two types of deprivation indicator which can be used. The first involves groups of people who are particularly vulnerable to poverty incidence, such as the unemployed, the semi-skilled and unskilled manual workers, and the underemployed small farmers which are subject to a two-level spatial analysis in the work of Nolan *et al.* (1994) on the Irish Republic. The second type of indicator involves outcome measures which are more direct expressions of actual deprivation, albeit somewhat restricted ones.

The term deprivation can be applied to a multitude of circumstances, encompassing social, physical and emotional aspects of life as well as the material deprivation with which this research is concerned. Moreover, an individual may be affected by one or more forms of material deprivation without necessarily experiencing income poverty. However, those who are subject to a very severe level of deprivation, or to deprivation on a number of fronts, are highly likely to suffer deprivation of income and resources (Townsend, 1987). For this reason, it is essential to examine different dimensions of deprivation and to pay particular attention to the existence of multiple deprivation.

Spatial deprivation analysis must be referenced to three separate dimensions of data, namely indicator variable, area base and time. The last of these presents no great problem, as our interest relates to the most recent datasets available, in this case the censuses carried out in both countries in 1991. The other two issues are more complex; while the selection of indicator variables and area bases for a single-state analysis is restricted by the limitations of one national dataset, cross-national analysis is constrained further by the degree of overlap between two separate datasets. The problems which are involved are summarised in Table 4.1.

Table 4.1a: Data Limitations Affecting Indicator Variables

	Single-State Analysis	**Cross-National Comparison**
Domain of Interest	National census questionnaire	Overlapping domains
Specific Variables	Definition + categories (i) collection (ii) release	Overlapping specifics
Tabulation of Variables	Form of release (i) univariate denominators (ii) cross-tabulation	Similarity of tables
Poverty Meaning of Variables	Representation of deprivation across different environments	Different national contexts
Composite Deprivation Indices	Multidimensional: limited only by domains	Highly limited by specifics

Table 4.1b: Scale Issues in Area Analysis

	Single-State Analysis	**Cross-National Comparison**
Constant Scale Issue	Variability in area-size	Difference in mean area-size
Settlement-Size Effect	Systematic mesh variation	Differences in mean area-size for each settlement band

SINGLE STATE DEPRIVATION MEASUREMENT

There is a large body of literature relating to the geographical analysis of deprivation using Census data in Great Britain, but rather less research has been undertaken on the topic in Ireland. Examples of some of the variables more commonly employed either as univariate indicators or as components of composite indices in the UK and the Republic of Ireland are presented in Table 4.2.

Table 4.2: Census-derived Variables Used as Deprivation Indicators in Great Britain and/or Ireland

Dimension	Variable
Employment	Unemployment
	Long-term unemployment
	Labour force participation
	Economic dependency
	Small scale farming activity
Education	Lack of formal qualifications
	Early school leaving
Social Class	Manual/unskilled workers
Housing Tenure	Owner-occupation
	Local authority housing
Car Ownership	Households without a car
Housing Conditions	Lack of central heating
	Poor amenities (no toilet, bath or shower)
	Overcrowding
	Rooms per person
Socio-Demographic	Children in low-earning households
	Lone parent households
	Elderly living alone
	Age dependency

Deprivation indicators may describe features of employment, occupation, educational attainment and social class, tenure, ownership of assets (homes and cars) and housing conditions. Other socio-demographic variables such as lone parenthood or elderly people living alone are also employed as indicators of "at-risk" populations. Univariate indicators are often combined into composite indices in an attempt to accommodate the multi-dimensional nature of deprivation, and so that overall deprivation rankings can be derived. These rankings may be employed to allocate financial resources; the £30 million paid annually to General Practitioners in the UK as a Deprivation Supplement is distributed on the basis of ward scores using the Underprivileged Areas Index (Jarman, 1983, 1984). Most of the indices in common use are based on *a priori* definitions of deprivation rather than being founded on empirical data (Noble *et al.*, 1995), and there is little agreement as to the most appropriate variables to use and the weightings to apply (Morris and Carstairs, 1991).

There are a number of problems inherent in the use of deprivation indicators which affect any study. The most significant of these are concerned firstly with the appropriateness of the choice of indicators; secondly, with the validity of comparing rural and urban conditions using the same indicators; and thirdly with the scale of analysis.

Representativeness of Indicators

The unemployment rate falls into the first of these categories, where the ability of the variable to adequately represent deprivation across different types of areas may be in question. Unemployment carries with it a high risk of poverty and material deprivation, particularly if periods of unemployment are frequent or prolonged, and unemployment is therefore a key variable both in the univariate analysis of deprivation and in the construction of composite deprivation indices. However, in certain types of areas (particularly remote rural ones), employment possibilities may be so scarce as to force much of the economically active population to move away from the area in order to find work. This type of selective regional out-migration therefore causes unemployment figures to drop, with the consequence that deprivation as measured by unemployment appears low; but despite the fact that such populations do not therefore experience high unemployment rates, it is also true that these areas are in fact too deprived to be able to sustain a normal proportion of economically active people. Thus, while the remaining population may or may not experience other forms of deprivation, it is clear that from a policy perspective the input of resources (in terms of investment and employment creation) would be necessary to prevent the continued depletion of the economically active population. This paradoxical situation must be taken account of by considering instead demographic or economic indicators such as economic or age dependency. Another limitation of this indicator is that unemployment figures do not reveal potential deprivation arising from underemployment, which is a particular risk for those engaged in small-scale farming (Williams, 1993).

Applicability to Urban and Rural Environments

An example of the problems of defining deprived areas across both urban and rural environments is illustrated by the variable relating to car ownership, which is an indicator widely used both as a univariate indicator and in composite spatial deprivation indices (e.g. Townsend, Phillimore and Beattie, 1988; Green, 1994; GAMMA, 1995); lack of a car is assumed to indicate relative material deprivation, and multiple car-ownership is held to indicate prosperity. It is evident that many individuals may well choose not to own a car for reasons other than purely financial pressures, and therefore the appropriateness of the variable may be questioned at the level of the individual household; however, when considered at area level, car ownership tends to differentiate well between poorer and wealthier areas. The validity of this indicator is more debatable when both urban and rural areas are included in the analysis, the need for personal transport being much greater in a remote rural area where the provision of public transport is low and service accessibility poor. Families which are already financially stretched may be forced to make further sacrifices to be able to afford the luxury of a car, thus although car ownership figures in a rural area may be higher than in an urban area of comparable average income, the remaining income available for other household needs may in fact be lower.

Scale of Analysis

It is well known that the scale at which spatial data is analysed influences the outcome of that analysis; at the smallest scale, that of the individual or the household, homogeneity of conditions is maximised, and the range of results will include many extreme values, while as the scale is enlarged, encompassing more and more diversity, conditions converge more towards an average state. It is therefore generally understood that in order to avoid biasing the results by scale-mixing among the set of areas to be analysed, comparative research should be conducted by adopting a constant scale level, usually defined in terms of units of approximately equal population counts. As the purpose of this research is to undertake analysis at the smallest scale possible within the constraints of data availability, thereby attempting to identify deprivation among small communities, the decision to undertake analysis at the level of the District Electoral Division in the Republic of Ireland, and the enumeration district in Northern Ireland, is not in question. However, a problem which is often overlooked, and which is of particular concern in this analysis, is that of a settlement-size bias, an issue which can give rise to a problem which has been labelled the "rank-order fallacy".

Briefly, this is the outcome of comparing spatially aggregated data across a range of settlement sizes, within which, because of the nature of residential segregation, the heterogeneity of socio-economic conditions varies widely. It is clear that conditions will be more homogeneous, and therefore tend more towards extreme values, if they are measured among a unit of population living in a large housing estate located in a city, than among the same number of people in a rural or small village environment which will encompass a larger number of "neighbourhoods" and wherein conditions will be more variable. This issue is not significant if the purpose of research is only to identify spatially extensive concentrations of deprivation; but in environments where poverty exists at a more micro-level, comparative analysis using a constant scale level can mean that urban deprivation blackspots dominate rank-order listings and deprived rural areas are overlooked when policies are devised or resources allocated, even though the level of poverty may be no less severe than that found in the cities. This issue is of particular significance in an Irish context, as highlighted by Moore's (1993) research in Northern Ireland. It is therefore evident that holding scale constant during analysis may not always be the best procedure, and that a different method of standardising scale may be required.

CROSS-NATIONAL DEPRIVATION MEASUREMENT

The problems which have been briefly described are further exacerbated by the need to compare conditions among two different political administrations. Differing current or historical national policies, for instance on the provision of public authority housing, and differing taxation systems which affect the price of consumer goods (of particular significance being the cost of owning a car) are likely to affect the comparability of deprivation indicators. Other factors which may influence the outcome of comparative analysis include the lack of co-ordination in the collection, aggregation and release of census data, with the result that the definition of important variables may differ between the two countries. Furthermore, many variables are released by only one of the census authorities. Approximately 9,000 variables, in 75 cross-tabulations, are available from Northern Ireland's small-area statistics, while the Republic's data consists of 1,750 counts — yet despite the greater overall detail present in the Northern Irish data, some key topics, covered by the Southern census, are not available for Northern Ireland.

Comparability of Indicators North and South

The choice of indicators is of course firstly constrained by the limited number of variables which are actually held in common. Figure 4.1 shows potential indicators which are released for each jurisdiction, illustrating the theoretical overlap. While several important types of data are held in common, it can also be seen that certain variables are not reproducible for both countries, a crucial example in an Irish context being the number of persons involved in small-scale farming activity. While the census statistics for the Republic differentiate farming activity on the basis of the acreage farmed, the UK census classifies farmers only as employers/managers or as working on their own account. No further detail can be acquired from the Northern Irish data on this topic, leaving no possibility to derive a truly comparable indicator of the potential deprivation due to agricultural underemployment.

Figure 4.1: Overlap of Deprivation Indicators for Northern Ireland and the Irish Republic

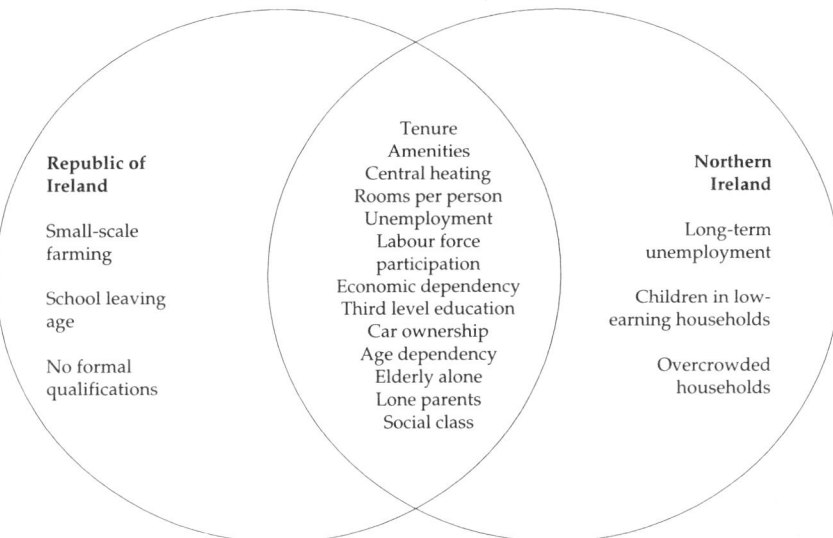

Another example is that of overcrowding, which is often defined in UK studies by the proportion of households which are crowded, whereas the closest variable available for the Republic of Ireland is the average number of rooms per person in the electoral unit. As would be expected, the two variables are strongly correlated — for instance, within Northern Ireland the correlation coefficient between the average number of rooms per person and the percentage of households with more than one person per room is $r = -0.86$ ($p<0.001$) at Enumeration District level; nonetheless, using an average measure for the areal unit means that the capacity of the variable to highlight areas with severe crowding problems is reduced if those households are located in the same spatial unit as much more prosperous, spacious households.

Comparatively little data is available at a small area level on educational attainment in Northern Ireland. While the Republic's census output includes eleven categories of attainment ranging from no formal qualifications to postgraduate degrees, the only small area data released for the Northern Irish population relates to third level qualifications — this is despite the fact that the census form actually categorises seven different qualification levels. That this data is not released means that comparative north/south research cannot incorporate indicators of low or no qualifications, an important measure of educational and labour market disadvantage. On the other hand, the Irish Republic's census releases this data only for the economically active, whereas the Northern Irish census categorises third-level education for both the active and inactive populations. In creating a variable to compare the proportion of the economically active population who have a third-level education, minor differences in the definitions of age groups exist — the Northern Irish data includes those from age 18 to pensionable age, while the data for the Republic covers all those over 15 who have completed their full-time education. Most significant, however, is the fact that the educational attainment indicator is so restricted by the limited overlap that it is only possible to produce a variable which is more a measure of labour market forces and of out-migration than of educational disadvantage.

Another issue concerns variables which are comparable in terms of definition, but which have overall national levels which differ so markedly between each country that in terms of socioeconomic deprivation they cannot be considered to be equivalent. Examples are housing tenure (publicly-owned and owner-occupied housing), household amenities such as the use of a bath or shower, and the presence of central heating. Table 4.3 presents national average figures for these indicators, while two of these variables are mapped in Figures 4.2 and 4.3, giving a rather conflicting impression of deprivation on each side of the border. It is clear that these variables cannot be used as the basis for comparison in a raw form unless they genuinely reflect deprivation levels equally.

Table 4.3: Indicators of Housing Deprivation — Percentage of Households Affected in Northern Ireland, the Republic of Ireland and the Border Regions

Region	Public Housing	Not Owner-Occupied	Without Bath/Shower	No Central Heating
Republic of Ireland	9.8	20.0	5.9	40.7
Northern Ireland	29.4	37.5	1.7	17.2
RI Border Region	7.2	15.9	9.0	45.8
NI Border Region	30.7	38.1	2.5	19.2

Figure 4.2: Percentage of Households in Public Ownership

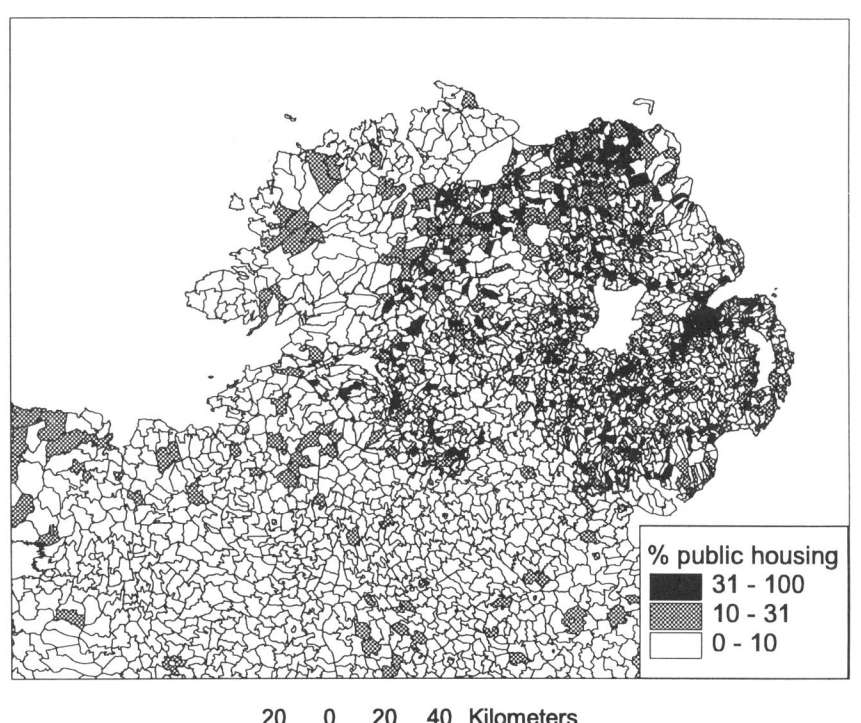

Figure 4.3: Percentage of Households Lacking a Bath or Shower

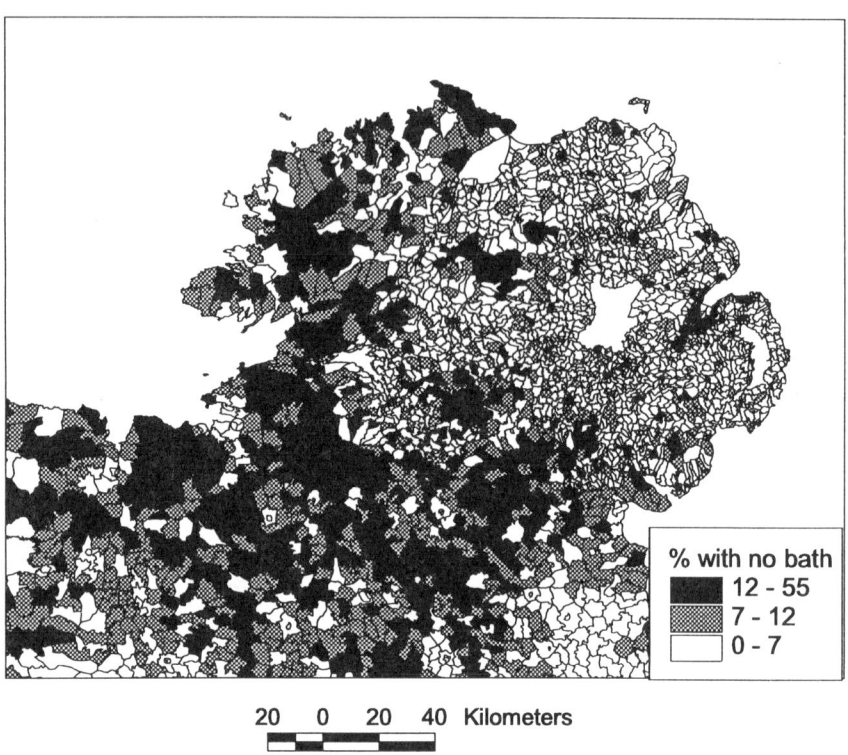

While standardisation of the area-level data within each country would remove the disparity, highlighting the most deprived areas within each state, evidently national comparisons would be invalid. It should be noted that some of the relationships which are found in the Republic between housing variables and other deprivation indicators are weaker or non-existent in the Northern data (for example, the correlation coefficient between central heating and car ownership in the Republic is 0.44, compared to only 0.07 in Northern Ireland) — it is likely that this is primarily a result of a large supply of relatively modern and well-maintained public sector housing in the North, affecting the relationship between housing quality and other features of socio-economic deprivation. Of course, the amenities indicators still have some value, because they indisputably demonstrate a form of housing deprivation, but the variables are not likely to imply the same levels of income deprivation in the two countries.

Evidently, the difficulties of finding variables which are suitable for a cross-national comparison particularly affect the prospect of applying composite deprivation indices to the Irish data. None of the standard deprivation indices used in the UK would be suitable for this purpose without some degree of modification.

Differences in Scale

Issues surrounding the scale of analysis have already been referred to in the context of national deprivation studies. A further consideration in a cross-national context is the influence of additional scale mixing, resulting from the combination of data from two different sets of administrative boundaries. The average population of a District Electoral Division is, at just over 1,000, more than twice that of an enumeration district, although the medians are much closer (at 490 and 426 respectively). In fact, the frequency distribution of population sizes in the Republic is strongly skewed by a relatively small number of highly populated DEDs; the largest DED population in the Republic is, at over 25,000, more than ten times that of the largest ED, giving rise to a danger of inappropriate scale mixing if all data is grouped and analysed together. Unfortunately, the difference in scale is not just confined to large cities; whereas the northern Irish towns are divided up into many enumeration districts, the majority of urban settlements in the Republic (other than the cities themselves) are composed of only one or two DEDs, with population sizes often several times greater than those normally found in urban EDs. The outcome of comparative urban analysis will therefore be that the more homogeneous spatial units in northern towns are likely to dominate joint north/south rank-order listings of urban areas.

This problem is illustrated in one of its most extreme forms in Figure 4.4, which maps male unemployment data for the similarly sized towns of Coleraine and Dundalk. Census data is released for over 30 separate urban enumeration districts in Coleraine, and

Figure 4.4: Impact of Urban Scale Differences in Cross-border Comparative Analysis: Male Unemployment Levels for (a) Coleraine, NI and (b) Dundalk, RoI

Coleraine Urban Area

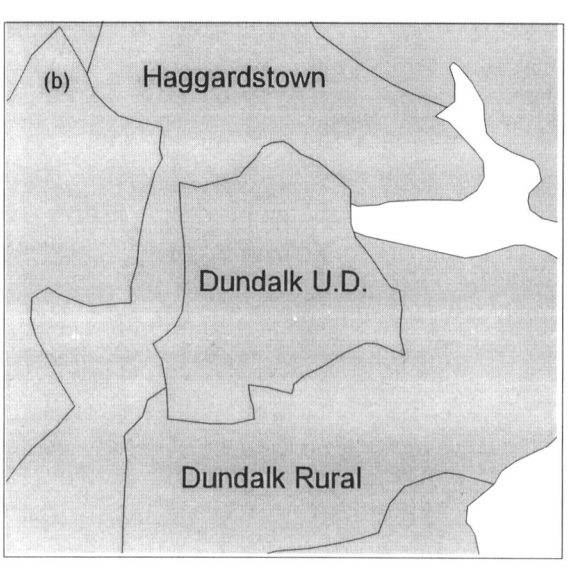

Dundalk Urban Area

within some of those which are located in public housing estates the unemployment rate exceeds 30 per cent, in one case reaching 48.5 per cent. Dundalk, on the other hand, is comprised of only one urban DED, for which the average unemployment figure is 28.9 per cent, 9.1 per cent higher than the average figure for Coleraine. Despite the fact that the overall unemployment rate is greater in Dundalk, and the distinct possibility that the unemployment level in the poorest parts of the town may rival or even exceed those of Coleraine's worst districts, rank-ordering which includes small-area data for both towns would obviously indicate conditions to be far worse in Coleraine.

Working within the Constraints of the Data

To undertake a cross-border deprivation analysis, then, it is necessary to work within the confines of a severely restricted set of comparable variables and to handle the data in such a way that the impact of scale differences is minimised. There is little that can be done to mitigate the problems which arise because of the limited overlap of the two datasets, other than to ensure that indicators chosen from overlapping domains are similar enough in their specific definitions that like is compared to like, not merely that similar dimensions are compared. This means having to dispense with indicators which are perfectly valid in a single-state study, an added consequence being that the more established deprivation indices cannot be applied. Experimentation with different compositions of indicators, and examination of relationships between variables in different environments is required to develop acceptable indicators and composite indices which are meaningful in an all-Ireland context.

The variation in the degree of homogeneity of conditions across different settlement sizes can most effectively be countered by generating rank-order listings separately for different environments (e.g. cities, small towns, rural areas) although if overall deprivation rankings are required, amalgamation of spatial units to achieve greater similarity in the level of homogeneity/ heterogeneity within the units may be necessary. While the scale problem in rural areas is relatively minor, the larger size of urban units in the Republic is of particular concern, as small areas of deprivation within many of these towns will be averaged out by more prosperous areas in the same unit, unlike the situation in the north where urban deprivation blackspots will be more visible. Amalgamation of units is, of course, a far from satisfactory solution to the problem; it is an unfortunate conclusion that the only way of putting comparative cross-border analysis on an equal footing is to cater to the lowest common denominator and thereby make data for both countries equally inadequate — and yet in the short-term, a loss of detail is necessary to prevent the rank-ordering of results from being biased by incompatibilities of data release.

The short-term solutions to these methodological problems can therefore be seen as an exercise in damage limitation. The longer-term solution lies in a thorough review by both governments of the content of data collected and released, and of the spatial aggregation of statistics, to allow for a greater degree of harmonisation of output. The demand from governments, academics and private ventures for comparable cross-national data will inevitably grow in the coming years, and rationalisation of the data should not be delayed — as it is, the length of time it takes to operationalise changes, the infrequency of censuses (especially in the north) and the lag-time prior to data release mean that it will be several years before any changes initiated now would come into effect.

CONCLUSION

There are two major sources of methodological problem in exploring the geography of poverty in Ireland. These involve the choice of indicator variable and the choice of area base. There exists a large literature in many countries dealing with the issue of the choice of indicators for expressing poverty and deprivation, but the choice of area has received considerably less attention, even from geographers whose *forte* this should be. There is a danger of analysts simply selecting the smallest geographical areas for which data are available and unthinkingly accepting a matrix of areas whose only common feature is a shared generic name, such as wards or enumeration districts. The result can be a very misleading set of empirical findings.

These observations apply to any analysis within a single political state. However, because partition has created two separate jurisdictions within Ireland, there are two distinct data regimes.

Therefore, there are major problems of cross-national comparison which can only be countered by administrative action. There is an urgent need to co-ordinate the national census operations, so that not only the questionnaires but the analysis, too, can be harmonised as much as possible. We are not suggesting that one should be a clone of the other, but surely deviations can be eliminated unless a good case can be made for maintaining a specific difference. With both common membership of the European Union and a shared international frontier, this seems an elementary need.

A common set of variables between the two national censuses is of little use, however, unless there is a closely matching spatial framework. The Republic of Ireland could do much to remedy this situation by releasing data for smaller areas equivalent in size to the enumeration districts of Northern Ireland. This would also solve another problem relating specifically to the Republic, for a spatial matrix of equal-sized areas is needed to replace the current massive variability of the District Electoral Divisions.

If governments are serious about tackling poverty and deprivation with an area-based approach, they need to provide a data-set which allows target districts to be selected in an objectively justifiable way rather than just as a response to lobby-group pressure on behalf of high-profile neighbourhoods. This applies to both within-state and cross-national action. It is in the cause of social justice, therefore, that we highlight some of the political efforts needed to solve the methodological problems focused on in this paper. It is not enough to have schemes and funding to relieve poverty — we must have action to improve the flow of information on where poverty is actually located.

NOTES

1. This chapter is based on research conducted as part of a joint project between the authors and Dr D.G. Pringle, National University of Ireland, Maynooth. Financial support for this project was provided by the Combat Poverty agency, Dublin and the Northern Ireland Voluntary Trust, Belfast.

REFERENCES

Bradford, M.G., Robson, B.T. and Tye, R. (1995) "Constructing an urban deprivation index: a way of meeting the need for flexibility", *Environment and Planning A*, 27, 519-533.

Callan, T. and Nolan, B. (1994) "The meaning and measurement of poverty," in Nolan, B. and Callan, T. (eds), *Poverty and Policy in Ireland*, Gill and Macmillan, Dublin, 9-26.

GAMMA (1995) *Limerick City APC Report*, Commissioned by ADM Ltd., Dublin.

Green, A. (1994) *The Geography of Poverty and Wealth*, Institute of Employment Research, University of Warwick.

Jarman, B. (1983) "Identification of underprivileged areas", *British Medical Journal*, 289, 1705-1709.

Jarman, B. (1984) "Underprivileged areas: validation and distribution of scores", *British Medical Journal*, 289, 1587-1592.

Moore, A.J. (1993) *General Practitioner Workload: a Socio-spatial Study in Northern Ireland*, Unpublished D.Phil thesis, University of Ulster, Coleraine.

Noble, M., Cheung, S.Y., Smith, G. and Smith, T. (1995) "Using census data to predict income support dependency", *Policy and Politics*, 23(4), 327-333.

Nolan, B., Whelan, C.T. and Williams, J. (1994) "Spatial aspects of poverty and disadvantage", in Nolan, B. and Callan, T. (eds), *Poverty and Policy in Ireland*, Gill and Macmillan, Dublin, 214-239.

Townsend, P. (1987) "Deprivation", *Journal of Social Policy*, 16(2), 125-146.

Townsend, P., Phillimore, P. and Beattie, A. (1988) *Health and Deprivation: Inequality and the North*, Croom Helm, London.

Williams, J. (1993) *Spatial Variations in Deprivation Surrogates*, The Economic and Social Research Institute, Dublin.

Part 2

RURAL POVERTY

5

RURAL POVERTY — A POLITICAL ECONOMY PERSPECTIVE

Hilary Tovey
Department of Sociology, Trinity College Dublin

INTRODUCTION

The focus of this chapter is on poverty as a *dynamic* condition, that is, as produced out of, maintained and/or changed by dynamic processes in society in general. It is only by thinking of poverty in this way that it makes any sense to speak of "rural poverty" — or indeed, "urban poverty". Fundamentally, these can be different from each other only as long as the rural is understood as the site of specific societal development processes which are not found, or not in the same way, in urban areas.

To understand *rural* poverty, in other words, we must start from an analysis of the processes of change and restructuring which are working through rural societies, particularly in the developed societies of the world, and try to achieve some insight into the differing effects which these have on the situation and life chances of different groups of rural people. These processes are in many ways distinctive and are not identical with processes of restructuring which are found in urban areas. Moreover, this approach implies that when we talk of rural poverty we are not concerned with the rural as space: space in itself does not generate either poverty or wealth. What does, is the way in which spaces are *used* in the process of development. Wealth, and poverty, are distributed across space as the result of societal — and increasingly, global — processes of investment, disinvestment, restructuring and economic reorganisation which take advantage of existing spatial differences in order to maximise profitability and security.

The uses made of space do not affect all of those living within particular spaces in the same way. Their impacts are experienced differently by different social classes, different age groups, different genders, and sometimes, different cultural groups. In rural areas, for example, both gender and age affect access to property or other material resources — perhaps less so than in the past but still to a marked degree; they also, usually in the reverse pattern, shape access to education and knowledge, to marketable skills and to cultural capital. Changes in how a rural area is used in a development process, for example from exploitation of agricultural resources to multinational-capital-based industrial production, may alter the relative worth of ownership of land versus ownership of marketable skills, and so alter the distribution of poverty along age and gender lines. This does not imply that space in itself has no bearing on the existence or experience of rural poverty, but only that approaches to rural poverty should not start with space but with *social groups*, and the social processes which produce these groups in a particular place and time. In other words, we are not concerned here with "poor places" but with "poor people". Poor people may be found in rich as well as poor places, and both sorts of place may be found in a rural setting.

WHAT IS RURAL POVERTY?

I do not intend to spend much time discussing different ways of defining poverty. This has already been done in other chapters in this volume, and there is also a quite lengthy discussion of how one should understand and conceptualise poverty, specifically when studying rural poverty, in the book edited by Chris Curtin *et al.*, *Poverty in Rural Ireland* (1996), which I draw on here as appropriate.

To state my own bias at the start, however, the approach to poverty which I find most acceptable is that which treats poverty as a relative, not an absolute condition (Curtin *et al.*, 1996, 5). Using a "poverty line" to locate people who are at a stated level (40 per cent, 50 per cent, or 60 per cent, depending on the choice of the researcher) below the average in their society, in terms of income or material wealth, is an example of a relativist approach to poverty. As this example suggests, relative definitions of poverty focus our attention on *social inequality* in a way other definitions of poverty do not — particularly inequalities in the distribution of or access to resources which significantly affect lifechances. Where, as is the case in most developed societies, poverty levels are not such as to (in general) threaten physical survival, this seems to me to be the most appropriate perspective to use. A relativist approach to poverty suggests that ultimately in looking for solutions to poverty we cannot avoid thinking about restructuring society to make it *more equal*. I argue below, in fact, that restructuring society so that existing resources are distributed more equally may be a more effective answer to the problem of poverty than trying to increase the absolute level of available resources through economic growth, in the hope that some of the increased wealth will eventually trickle down to the poor.

On Defining the Rural

It is less easy to avoid offering some discussion of the meaning of "rural", given the arguments outlined at the start of this chapter. The rural is a particularly difficult concept to lay hold of because it appears in public discourse in at least two forms. There is the rural as the object of social scientific research; and the rural as the subject of ideology. There is also a third way in which the rural appears frequently in public discourse — that is as the object of policy or policy interventions; this usually combines scientific and ideological understandings of the term in more or less unreflective ways.

The rural as the object of research is assumed to have some sort of objective existence, which can be recognised and categorised. Rural areas are those which have a certain (low) density of population per square mile, perhaps; or they are areas where forms of natural resource exploitation predominate in the local economy; or (increasingly) they are areas characterised by the presence of "nature" or of "environmental" resources. There are of course major problems, particularly between different disciplines (such as sociology and geography), in reaching agreement on what the objective, defining features of "the rural" as opposed to "the urban" actually are. There is also the problem that differentiation within the rural, between different localities and different regions of the country, is historically well established and is becoming more pronounced over time. This raises the question whether there is anything to be gained, from a research point of view, from lumping such a range of very different socioeconomic and spatial entities together in the single category of "rural", or whether in fact this does not end up concealing more than it reveals. A number of books on rural change in recent years suggest that differences between rural areas are now as great as or greater than any differences between rural and urban.

Most research on rural Ireland which constitutes the rural as an objective entity has been carried out from the modernist point of view, which assumes that modernity is urban and industrial and that therefore rural areas are always in some way a problem for society. They are the location of tradition and need to be modernised, they exhibit poor average incomes per head of population or poor levels of education, they experience outmigration, and so on. This way of thinking about the rural encourages discussions of rural poverty which treat that as a feature of areas, rather than of people. It pushes us towards looking for indicators which show which areas are poor, and which are relatively well off. It also encourages researchers to look within the rural areas themselves for the causes of their poverty. It does not encourage us to ask whether it might be the "developed", urbanised areas and their activities which are the problem, in the sense that it is there that we should look to find out what is generating and reproducing poverty over time among specific rural groups.

Alternatively, the rural appears in discourse as the subject of ideology. By this I mean that societies generally attach specific symbolic meanings and evaluations to the countryside, to rural living and to rural activities such as food production which shape and influence how those who live there experience their lives. We might go further and say, following Mormont (1990), that the distinction between "rural " and "urban" which gives the rural any content it may have is a product of the ideas which exist in society and which are mobilised by specific social groups for their own ends. When actors within the society cease to find "rural" or "urban" useful as mobilising concepts then there will be no meaningful rural-urban difference in the society.

As the subject of ideology, the rural does not necessarily appear in discourse as "a problem" at all. Indeed quite the opposite — it is a striking feature of most modern, urbanised, industrialised societies that they are characterised by ideologies which elevate rural life and the rural environment into potent symbols of goodness. It is this which explains the fact, noted by a number of researchers on rural poverty in Britain, that some rural groups who in terms of objective indicators are poor do not describe themselves as poor or do not regard their poverty as an intolerable problem. The subjective meaning of poverty to the poor — and to the better off — is strongly shaped by the culture and the ideologies present in their society. But of course these differ from society to society (and between different social classes within societies), and we should not assume that because in the advanced social cores of Europe rural living is seen as desirable and good, this is also the case in the less developed and more peripheral countries. Ireland has experienced a massive movement of people over two centuries, and not only for economic reasons, from rural areas to cities, usually abroad. Those who remain in the countryside were told repeatedly that they are the ones who lack intelligence or initiative, and have seen their way of life systematically devalued and delegitimised (Brody, 1973). Rural poverty in Ireland is quite unlikely to be experienced in the same way as rural poverty in England.

THE PRODUCTION OF POVERTY — SOCIAL PROCESSES IN RURAL AREAS

This chapter began with the claim that rural areas should be approached as the site of specific and distinctive economic and social processes. To that extent, it does adopt an "objective" definition of the rural. To develop that further, I argue here that we need to start from a view of "rural Ireland" as the spatial and social manifestation of global capitalist development. In this sense both rural and urban areas are shaped by the same broad developmental processes, national and particularly international; but the process has different outcomes for different sorts of location. This is primarily because of the way in which capitalism operates through a spatial division of labour which locates different types of economic activity in different types of space. In thinking about rural poverty, then, our main focus should be on how socioeconomic restructuring within a global setting distributes and redistributes resources and opportunities between different social groups.

How then is the rural distinctive? *Poverty in Rural Ireland* put forward a case for arguing that the most important marker of rural areas, which gives them what common features they possess, is the presence in them of a history of *agricultural production* (1996, 11). The reference to a "history" of agriculture is important; we cannot assume that agriculture is still the dominant production process or employer of labour in all or even most rural areas in contemporary Ireland. Industrial and service occupations are replacing farming as the main occupation of a majority of rural inhabitants in many areas; many previously agricultural areas are characterised by population loss as people move into or near towns to live; in some areas, tourism is beginning to replace agriculture as the main user of rural space and natural resources. Nevertheless both the spatial and social characteristics of rural populations are still best understood as the result of the presence, in the locality, of an agricultural economy in at least the recent past.

If this argument is accepted, the place to start in a study of the distribution and dynamics of rural poverty is the organisation of agriculture and food production. Irish agriculture currently is characterised by two particularly striking features. One is the sharp decline in the number of "farmers" (though the decline in the number of landholders appears to have been much less) which accompanies a pattern of increasing concentration of production and output. The NESC Report *New Approaches to Rural Development* estimated that around 60 per cent of total farm output and 40 per cent of total land farmed is found on around 20 per cent of farms (Commins and Keane, 1994, 45); such farms have farm incomes which are around three times the average farm income in Ireland (Frawley and Commins, 1996, 23). For the other 80 per cent of farm households, livelihoods have increasingly to be sought off the farm, in employment or in receipt of state benefits. The second is the pattern of regional shift in agriculture, also well described in the NESC Report and in Commins (1996): agricultural production which used to be dispersed across the island has become concentrated on larger commercial farms in the south and east. Again, landholders outside these areas are increasingly marginal to agriculture and food production. To explain these sorts of changes in contemporary Irish agriculture we have to see it as part of and shaped by an *integrated food system* (Goodman, 1991; McMichael, 1994) which operates increasingly on an international or global level.

Concentration and Vertical Control in the Global Food System

We can distinguish two main types of strategic actors involved in the global food system: transnational corporations (TNCs) and national states. The food industry sector, globally, appears to be more tightly concentrated in the hands of a few very large corporations than almost any other sector of production (Bonanno *et al.*, 1994; Tovey, 1991). Goodman *et al.* (1987) suggest that the TNCs involved increasingly tend to divide into one or other of two types. Some of them are involved in multiple lines of activity, and are particularly likely to have investments in the chemicals and biochemicals sector; this is linked to their interest in finding ways of producing food industrially, using biochemical technologies and materials (a process Goodman *et al.* call "substitutionism" because it involves substituting industrially produced food ingredients for those produced on the farm). Others are rooted in an area of the conventional food system (transportation or processing, for example) from which they have expanded to develop integrated chains of interrelated activities often around a specific food item such as lettuce or chicken. Heffernan (1984) and Heffernan and Constance (1994) describe situations where companies which started by exporting and shipping US surplus grain moved back from that to control virtually the entire food production system associated with grain-fed poultry or beef — from (in the case of poultry) the chicken hatcheries through to the production of consumer-ready chicken pieces in franchised fast food outlets.

Globally, the capitalist food and agriculture system is organised around an international division of labour intended to make maximum profit, not only from regional differences in the availability and cost of labour, but also from variations in climatic and time zone differences in different parts of the world. For example, Mexican peasant holdings are used to produce strawberries and other luxury fruit and vegetables for an out-of-season North American consumer market, to which they can be flown within hours of picking. The food system works primarily on the basis of expanded control by capital over on-farm production processes, with corporations usually (though not always) more interested in getting access to and control over family farm labour, for example through production contracts, than in buying up land and engaging directly in production with an employed labour force. So-called "family farming" appears to be particularly suited to this form of control, in which capital allies with science and technology to supply the inputs and the rules for managing on-farm production needed to produce output at the levels of quality and price stipulated in the production contracts.

The result of such developments in the global food system is that increasingly, farmers are faced with the option of either becoming "vertically integrated", generally as powerless and dependent participants, into TNC-managed industrial food production systems, or being bypassed and marginalised from commercial agriculture. A recent study of organic farming in California shows that even in the case of non-conventional food production, where we would expect strong resistance to capitalist incorporation, trends towards integration and incorporation of the sort suggested above are nevertheless becoming apparent (Buck *et al.*, 1997). The farmers who are targeted by the food industry for incorporation are often the relatively well developed, intensive producers, while those who do not possess the capital or skills required become increasingly externalised from the food system. Irish agriculture has been to some extent protected from the impact of the global food system by the strong presence of farmers' co-operatives in the food processing and distribution areas — at least in dairy farming, this is obviously less true of beef — but as the co-operatives themselves become more concentrated into a few very big organisations, and as they transform themselves more and more into conventional stock-market listed corporations, this protection diminishes. Already it is becoming apparent that Irish milk producers are in competition with milk producers in other parts of the world as suppliers of the materials which Irish multinational food companies process and distribute, and which they are increasingly able to source on a global basis.

The other strategic actors in the global food system which we mentioned were national states. From the 1950s to the 1970s, most states managed their food sector and farm populations through the use of production and export subsidies and import controls as needed. The USA, as the increasingly dominant state in the international political order, also subsidised exports in order to try to restructure international trade and production in their favour — as later did the EU as it too began to experience surplus production. In recent decades this has increasingly led to the incorporation of agricultural production and trade into global trade negotiations such as under GATT. The fortunes of individual farm families in Ireland, and their chances of reproducing themselves as farmers into the next generation, are increasingly determined in an international political arena which is dominated by considerations of geopolitics and national or regional interest.

During this period, in most developed countries, there has also been a fundamental shift in political power and influence away from the "food-raising" or farming sector to the food-input and food processing and marketing sectors of the food system. This switch has been quite evident (although surprisingly little discussed) in Irish state policy for agriculture over the past decade (Tovey, 1991), as state resources for farm development are increasingly tightly targeted at those farmers who are likely to be able to produce food in ways which suit the food industry (for example, those who can guarantee standardisation of quality, or year-round maintenance of supply) rather than on farmers who are in need of development aid. Paralleling this at a global level, there has been a shift in power from national states to the emergent TNC food giants, who, as Curtin *et al.* point out, "increasingly experienced national regulatory frameworks as an obstacle to further integration of a potentially global agro-food sector" (1996, 19) and who have been instrumental in moving GATT talks towards agreement on dismantling such frameworks.

These processes within the international food system mean that, as long as national states seek to remain integrated into the global economy, it is less and less an option for them to try to address poverty among farmers through subsidies for agricultural production. It is now clear that, in any case, income supports via production subsidies in both the EU and USA were not a solution to farm poverty; they had the effect of widening the gap between larger and smaller producers (Commins, 1996), while reproducing small farmers at an income level where they were still vulnerable to pressures to marginalisation and economic exclusion.

Increasingly states are looking for alternative ways of responding. Within the EU, farm income support policy appears to be moving towards direct income supports which are dissociated entirely from food production, and also, to some extent, returning these supports to national management and control. Given the historical experience of unemployment benefit payments among non-farm groups in Ireland, and the concentration of households headed by an unemployed person among those currently in poverty (Callan et al., 1989; 1994), this does not bode well as a solution to rural poverty. Other responses within the EU have been to encourage diversification by farm populations into other, non-food uses of rural and natural resources, and to support part-time farming. Non-farm incomes have now become essential for many farm households to keep them above the poverty line (Commins and Keane, 1994; Frawley and Commins, 1996). Both policy responses further encourage the emergence of a heavily concentrated, industrialised and vertically integrated form of farming among those landholders who remain in full-time food production.

Industrialisation and Urbanisation in Rural Ireland

The growth, over the past half-century or so, of a globalised food and farming system in the form described above has occurred in parallel with a global restructuring of manufacturing industry. Again, this involved implementing a division of labour based on separating the location where production takes place from the centres of technical, financial and managerial control of the production process, which is very similar to what has been happening in farming with vertical integration into industrial food systems. While the managerial activities in manufacturing industry generally remained in or moved into positions closer to the socio-economic cores, the production activities, especially those which were less skilled, were often dispersed as much as possible into peripheral countries or peripheral regions of core areas.

In Ireland, as elsewhere, the contraction of farming in much of the country from the late 1960s on made its rural areas particularly attractive to many TNCs as locations into which to disperse the less-skilled but more labour-intensive stages of the production process: a workforce was becoming available which was dependent on off-farm income, with little history of unionisation or organisation, and which could be employed relatively cheaply. This was particularly the case when part-time factory work was involved, when low rates of pay or other benefits were rationalised on the grounds that the worker had another source of income and that the factory was deliberately accommodating this in its work arrangements.

Curtin *et al.* (1996, 23) claim that while Ireland suffered all the common ill-effects associated with a strategy of development through foreign investment,

> possibly the worst effects . . . were . . . that these companies represented highly mobile investments. As international markets fluctuate, companies decide at the stroke of a pen to open and close factories, with often devastating consequences for the towns and villages in which they had settled.

In more rural locations in particular, branch plants of multinational corporations were often the only significant employer in the locality. Attracting considerable proportions of the local workforce into the factory, they had the effect of encouraging people to discontinue their previous, more marginal economic activities. When such factories subsequently closed, the local economy could be left in crisis as it was often impossible to re-establish previous economic activities. Emigrant farm children, for example, attracted back into their home area by the prospect of industrial employment were often given sites on the family land to build a house, thereby making an already marginal agricultural holding completely unviable (Duggan, 1994).

The retraction in farming and the withdrawal of agriculture as the main user of the countryside has also opened it up to new forms of "urbanisation". These include the movement in of urban settlers, as permanent residents in the dormitory towns, for example, which have developed in proximity to most of the larger cities and particularly down the east coast, and/or as holiday home owners. Such changes in population patterns bring new opportunities for wealth to some rural residents — those with land to sell, people working in the construction industry, local shopkeepers and so on. They also bring new reasons for poverty, for example if they put upward pressures on local land or housing prices, or when the sale of farm land (particularly in the eastern counties) for housing development also means the end of employment for agricultural labourers.

But "urbanisation" also includes less visible changes than these. It includes the emergence of demands among an increasingly urban-based population that the countryside should be made available for alternative uses besides that of food production; the emergence of new understandings of what "the rural" means (open space, for example, rather than a type of society); the emergence of new ideologies (for example, of heritage conservation or of conservationist environmentalism) which impact both directly and indirectly on rural people by imposing new controls on what they can do with their resources, as well as opening up new opportunities. Increasingly, there is a tendency for the rural to become the battleground on which contests between opposing ideologies are fought out between essentially urban protagonists, contributing to a situation in which rural people themselves seem to have disappeared from the discussion (Tovey, 1994).

New Forms of Rural Resource Use

The concentration of agriculture in certain regions and localities of the country, and amongst certain types of farmers only, has created space for a range of rural resource uses other than agriculture. These new or not-so-new interventions into the rural economy range from the material exploitation of nature in the form of mining, forestry or fishfarming, to its cultural exploitation in rural tourism and heritage commoditisation, through to environmental conservation in the form of national parks, Natural Heritage Areas, the Rural Environmental Protection Scheme, and so on. In most cases — tourism is a particular example — they are seen as part of a strategy for rural diversification and integrated rural development which will enhance wealth generation and increase employment in the localities concerned. However, it is important to recognise that the gains from such rural resource development do not automatically go to rural people, or within rural populations to those groups most in need of them. Resource development may produce changes in the ownership of the rural environment, or in the forms through which it is worked, which benefit some groups far more than others (Tovey 1996, 129).

Rural resource development cannot be explained simply by focussing on the efforts made by local groups, or even by state planners, to enhance the livelihood chances of rural populations. Marsden *et al.* (1993) suggest that for a number of reasons there has been a renewed interest in the economic potential of rural (as opposed to agricultural) resources at the level of the global economic system. The economic recession of the 1980s produced a crisis of profitability for capitalism, which encouraged the holders of large capital in Europe and internationally to look for new opportunities for investment, and to reassess resources and spaces which were previously thought of as unprofitable or unproductive. In this context, rural space took on a new economic importance. No longer just places where workforces attractive to transnational corporations could be found, rural areas are also and increasingly the site of resources required by, or more easily exploitable as a result of, advances in science and technology; and they are becoming the site of consumption goods (living space, a clean environment, recreation, education) sought by an increasingly affluent urban middle class.

This suggests that there are distinct characteristics of rural space which are relevant to an understanding of rural poverty. But what is being argued here is that these characteristics are not "given" features of the rural — they appear because of the way in which economic or technological developments open up new ways of using rural areas or generate new demands for the resources they hold. It is not the rural space in itself which creates poverty, or wealth. Besides that, of course, as Marsden *et al.* (1993) emphasise, the processes of change in rural areas which result from the revaluing of rural spaces are very uneven. Some localities become incorporated into an international tourist, mining or fish-farming economy; others attract national or regional business investment; others again undergo new forms of state management and regulation, particularly in relation to the use of nature; and others, finally, become more marginal to national economic life than they were before.

In Ireland, most discussions of developments in rural resource exploitation assess these from the point of view of the contributions they make to increasing *national* wealth. However, from the point of view of understanding and addressing rural poverty, this is inadequate. What is important, as argued earlier in this chapter, is the way in which restructuring in the rural economy redistributes resources and life chances among rural people. The focus should therefore be on two issues in particular: (a) the impact of new forms of resource use on the opportunities of rural people, particularly the poorest among these, to gain a livelihood through working on indigenous resources; and (b) the opportunities they offer for external or foreign capital to penetrate into the localities. In an earlier discussion (Tovey, 1996), I argue that there is considerable evidence to support the conclusion that the two are not compatible: as external capital investment in rural areas rises, in the absence of a deliberate state policy to ensure otherwise the exclusion of the poor from secure local livelihoods is increased.

This reminds us again how weak the relationship is between the wealth or poverty of *a rural space* and the wealth or poverty of *a rural social group*. Use of a space to generate wealth (in mining, in tourism) does not necessarily make all those who live in it rich, or even better off than they were. In some cases, we might say (and both mining and some types of tourism provide examples of this) that to make money from the resources of a rural area, many if not all of those who live in it need to be impoverished. Curtin and Shields' (1988) account of how the development of lead and zinc mining in Tynagh required some local farmers to undergo loss of legal control over the land they owned, and loss of income from carrying on farming on that land, is a salutary reminder of this.

RURAL DEVELOPMENT AS A SOLUTION TO RURAL POVERTY

What, if anything, does the above discussion suggest about finding solutions to rural poverty? The first question we might ask is whether the fact that rural areas are subjected to some fairly specific processes of economic development and restructuring, different to those generally found in towns and cities, is relevant at all for solving poverty. It can be argued that even if the processes producing and distributing poverty are different, the needed solutions are much the same in each case.

There seems to me to be a good deal of sense in this position. Kautsky's famous Marxist analysis, at the turn of the century, of the problems of agriculture in the developing capitalist economy of Germany (*The Agrarian Question*, 1899), argued that what happened to agriculture in modern societies was increasingly the result of the changes occurring in the advanced, urban industrial sectors. Roughly put, he said that the engine of change in capitalist society is to be found in industry — agriculture, and with it rural populations, are merely so many carriages dragged along in its wake. It is industrial development which generates the new knowledges and technological advances, the mechanisation of work and the reorganisation of farm labour processes, the new ideologies of what farming is for (particularly, for producing cheap and reliable food for urban workers and consumers), which we see working themselves out in social changes in the contemporary countryside. The argument made here is quite similar, and its implication is clear: attempts to tackle rural poverty which do not also include struggles to change the broader economic systems in which agriculture and rural society generally are enmeshed are likely to have very little success. The problem of rural poverty, in this perspective, is a problem of modern society itself.

This might lead us to question the rebirth of interest in "localism" as a solution to rural poverty — local diagnoses of problems, local collective action to change them. As manifested in a plethora of recent projects for rural development, such as the Pilot Area Programme for Integrated Rural Development, the Operational Programme for Rural Development, LEADER, Area-Based (PESP) Partnerships, the County Enterprise Boards, the Community Development Programme, FORUM and others, local or spatial approaches to tackling rural poverty appear to be generated at least as much out of ideological perceptions of the rural as out of an objective analysis of the dynamic processes producing rural poverty. The NESC Report on *New Approaches to Rural Development* endorsed the use of territorial approaches to rural development, but even as it did so, it warned that "the emphasis on a territorial or area-based approach should not be allowed to exclude consideration of a sectoral policy" (1994, 19). Curtin *et al.* point out that not only may rural people interpret their material situation differently from urban people, they also

> may be subjected to rather specific understandings of how to deal with poverty, which are held by the leaders of rural society, by the makers of rural policy — themselves probably not "rural" — or by other groups in Ireland (1996, p. 11).

These include the assumption that there is something called "community" which persists particularly strongly in rural areas, which everyone is equally committed to, and which is a basis for mobilising collective action; and the assumption (which would hardly be made about the urban poor) that the rural poor should be mobilised along lines of territoriality rather than class — that is, as people who live in a particular area, rather than as people who share a common situation of economic and social exclusion. The imposition of these assumptions and expectations on the rural poor could be seen as *part of the problem which they face*, as much as part of the solution.

In a context where the rural areas of an already economically dependent small country, on the periphery (in terms of power as well as of space) of Europe, are so constrained in what they can do by the near-anonymous operations of an international capitalist economy, the emphasis in addressing poverty should be, as I have already suggested, at least as much on increasing equality as on achieving development. A recent analysis of EU regional development strategy (Scott, 1995) shows the extent to which a commitment to spatial equalisation, without a corresponding concern for social equality, can intensify the possibilities for social and economic (and cultural) exclusion of the less well off. Curtin *et al.* put forward a similar position:

> From the standpoint of the rural poor, intervention to equalise the distribution of existing resources, or to prevent or reduce the marginalisation of specific groups from the available opportunities, may sometimes be more beneficial

than trying to increase the overall level of wealth generated within a locality or region (Curtin *et al.*, 1996, p. 3).

Strategies to Increase Social Equality within the Rural

There are alternative ways of realising this sort of strategy, which are not necessarily mutually exclusive. One accepts that conventional rural "development" policies are unlikely to be abandoned, but calls for these to be accompanied by new and expanded intervention by the state. We have to recognise that the capacity of state policy to reverse undesirable outcomes of macro economic processes (while leaving those processes themselves in place) appears to be quite limited; nevertheless, if we are committed at all to a belief that social structures and relationships are not entirely economically determined, or that action can make a difference, we should consider very carefully what sorts of interventions are likely to be most useful.

What is being suggested here is that rural areas might benefit more, not from increases in specifically rural anti-poverty policy strategies, but from expanded and new forms of *equality policy*. These could include redistributive effects of state taxation and social services policy, for example, something which is relevant to both rural and urban people; but equality policies can also be tailored to specific rural processes, and attempt to implement forms of positive discrimination, suitable to rural and agricultural situations, of the sort pioneered by women's and minority movements internationally.

One possibility here is that the state might return to attempts to improve the access to land of smaller farmers, and of people from non-farming backgrounds who want to enter farming, which were part of the history of Irish agrarian policy for over a century, up to the disbandment of the Land Commission in 1980. The methods used to intervene in land markets today may need to be different, but the principles of intervention could be very similar. Another possibility is for state policy to discriminate positively, in the distribution of financial and of technical assistance and support, in favour of those types of farming which are most accessible for smaller landholders. Intensive promotion of organic farming might be a case in point — not just on environmental grounds (which would justify financial supports to large-scale organic producers as well as small) but on the grounds that this is one type of farming which, by revaluing food as a consumption good in an over-industrialised world, has the potential to return a good income to the small, extensive producer.

A second strategy for increasing equality within rural society calls for less, rather than more, integration of rural localities into national and global capitalism. Conventional understandings of rural and regional development are that to increase wealth, the economic structures and productive activities of "lagging" areas need to be opened up to and made available for incorporation into international capitalist relations, even if this produces temporary dislocation and loss of some livelihoods at the local level. Alternative theories argue that the displacement and inequality fostered by capitalism is neither temporary nor accidental, therefore what poor people need is not greater integration into it, but greater insulation and protection from it. This would suggest that an anti-poverty strategy for the rural should deliberately set out to support and strengthen institutions and movements which help to disengage local disadvantaged groups and economies from domination by capitalism — the credit union movement, a revitalised co-operative movement, movements to create local monetary independence such as LETS (Local Exchange and Trading System) (Douthwaite, 1996), and so on.

Reference to the women's movement, above, also suggests that we might consider whether in fact it is *more politics* which the poor need, rather than more policy. Politics is used here not in the sense of party politics but in the sense of engagement with power structures, or political mobilisation. It implies that the poor — specifically the rural poor if that categorisation made most sense to them, but this would need to be investigated — are actively encouraged to develop an understanding of their own situation and interests, and a "project" (Peillon, 1982) for the future which they can try to realise. Political mobilisation is, alongside cultural mobilisation, one of the few ways in which excluded and disadvantaged groups in society can empower themselves and develop a capacity to act. Combating rural poverty through more politics would require a close examination of the different "representative organisations" operating in rural society — farmers' organisations, community and development groups, local residents' associations and so on — to establish the extent to which they offer the poor an opportunity to address their own situation and to develop a collective movement for change. Most such organisations in rural Ireland are not class-based, and membership of the poor within them may well be one element in helping to keep them poor.

"Sustainable Development" — A Solution to Rural Poverty?
There is currently an increasing interest in the idea that solutions to rural poverty might be found in the theory and practice of "sustainable development". This is an intriguing and attractive idea; but we need to ask whether sustainable development, as the concept is understood at present in Ireland at least, does not prioritise other concerns (particularly environmental ones) over poverty and social inequality, to a point where it may in fact be an additional poverty-generating element in rural areas.

The past decade or so has been marked by the emergence of new interests around nature and the rural environment, particularly among urban groups. At the same time as economic activity in rural areas increasingly results from intervention by external interests, so "rural Ireland" increasingly becomes the object of competing claims to symbolic ownership by external groups concerned primarily with consumption of the rural environment. Among the most public and influential of the new claimants to rural Ireland is the environmental movement. Because environmentalism operates largely through a discourse of "global" environmental problems, it appeals to interests which seem to be held in common and equally shared by everyone. It is easy to assume that questions of inequality simply do not arise; obviously we all share the desire that "our" planet should survive. Nevertheless, I want to argue that there are important questions to be asked about whether the concerns of environmentalists for protection and "management" of the rural environment are compatible with the concerns of the rural poor for a better livelihood.

Most environmental discourse in Ireland has been dominated by conservationist ideas, which focus on *the rural as nature* and often regard rural people as external to and an intrusion on that nature; this is evident, for example, in the "wilderness" Model of National Park management which has been adopted by the Office of Public Works and which contrasts so strongly with, for example, French understandings of managing a regional park (Dwyer, 1991). Conservationist interests in Europe, whose rise in influence coincided with a fiscal crisis in the EU over continuing supports to farming, have been able to have a number of measures introduced which reduce or severely constrain the uses which farmers — particularly the poorest farmers in the poorest farming regions of Ireland — can make of their land to generate an income from producing food. The recently introduced system of National Heritage Area designation, and its incorporation into the Rural Environmental Protection Scheme for farmers, are two examples of the way in which conservationism prioritises the needs of "nature" (as interpreted by scientific and other accredited experts) over the needs of rural resource users.

Against this background, the movement towards a discourse of sustainable development, amongst politicians as well as environmentalists, seems to offer a welcome way of reconciling the interests of both the conservationists and the rural poor. A commitment to "sustainable development" seems to imply acceptance that rural areas need economic activities and that demands for conservation should not be allowed to prohibit this or make it entirely unviable; at the same time, it insists that no development should be allowed to take place which would permanently or significantly damage the rural environment. This in itself can be seen as in the interest of the rural poor who (as small food producers or tourism hosts for example) are often the most dependent of all rural residents on access to undegraded natural resources.

However, doubts remain about whether the way "sustainable development" in being understood in practice in Ireland allows it to incorporate any concern about addressing rural poverty. The most widely quoted definition of sustainable development is a selective quotation from the 1987 Report of the Brundtland Commission, which describes it as "development which meets the needs of the present without compromising the ability of future generations to meet their needs". This focuses attention on the way in which selfish abuse of resources by the present generation can rob our children and grandchildren of a reasonable standard of living. The fact that Brundtland also emphasises equity *within the same generation* — particularly (but not only) equity between the present populations of the developed and the undeveloped worlds — is largely ignored in Irish understandings of the concept.

Second, "sustainability" tends to be understood almost exclusively in terms of physical and ecological or else economic sustainability. Bord Fáilte's *Tourism Development Plan for 1994-99* puts a considerable emphasis on the importance of sustainable tourism for Ireland, for example, but defines this as "sustainable in terms of a quality environment . . . in terms of developing profitable enterprises [and] . . . in terms of enduring job creation" (1994, 1). What we might call "social sustainability" — which is concerned with things like the types of jobs created, the rate of pay of workers and the sorts of working conditions offered them, and the impact on the broader local society in destroying or transforming access by the rural poor to local resources, etc. — is not considered at all. The document *Sustainable Development — A Strategy for Ireland*, produced by the Department of the Environment in 1997 appears to have no conception of "society" at all, let alone of "rural society", but understands the world as made up of physical space and economic activities only. It understands sustainable development in Irish agriculture in terms of the production of "quality food" in a "quality environment". Who is to produce this food and who is to be marginalised as a food producer, and what treatment is to be meted out to farm labourers or farm animals in the course of its production, are issues not addressed. In summary, "sustainable development" has been, to date, incorporated into Irish policy discourse in terms which are almost entirely technological and technical, and which disregard the social reasons (precisely, the inequalities existing between the poor and the rich of the world which makes a flat refusal of any further economic growth or development impossible) why the concept was introduced in the first place.

There is, finally, a growing consensus in the international literature on sustainable development that attempts to implement it can not be either localised or technical only; sustainable development, if it is to be taken seriously, requires large-scale changes in the *social and political organisation* of society. Consider again the case of tourism: as a global business, tourism is evidently increasingly unsustainable, in terms alone of the huge numbers of people who are carried about the world on so many different forms of transport, using up valuable finite resources of fuels and materials. It is not possible to make tourism "more sustainable" by encouraging the development of agri- and niche rather than mass forms of tourism in local rural areas of Ireland. The most this can do is not increase the current unsustainability of world tourism too much. To make tourism really more sustainable, we would need to alter the patterns of recreation and consumption which characterise the "developed" industrialised world as a whole, and the beliefs about individual rights to self-realisation and self-expression through the exercise of consumer choice, which underpin these.

Sustainable development is not implemented just by developing ways of measuring environmental impacts to ensure that resources are not becoming depleted or that wastes are not too large to be assimilated by local ecosystems. Institutional and political structures also need to be changed, so that they start to enforce or reinforce sustainable practices. Again we find that we have to think in terms of major social change, not just in terms of setting limits on resource use. But if understanding sustainable development purely in technical terms leaves it unable to make an effective impact on environmental resource problems, how much less likely is it to offer ways of including, empowering and *sustaining* the rural poor?

CONCLUSION

The processes which are responsible for producing rural and agricultural poverty are not, in the main, located in rural areas. They are processes central to the contemporary industrial capitalist world, and are found close to the cores of global capitalism. There are, as I have argued above, specific elements of these processes which shape agriculture and food production in rather specific ways (for example, reproducing family or household enterprises as key elements in an integrated industrial food system, and concentrating farm production according to particular geographical patterns) and so give rise to new uses of rural labour and of rural spaces. But this does not mean that the problems of rural areas can be solved by working for change within rural areas themselves.

The perspective developed in this chapter is one which treats rural society as simultaneously less separate from, and more distinctive from, urban society than most other theoretical perspectives do. A political economy approach to poverty has another very important feature too: it shows us that in confronting poverty and inequality in the countryside (as much as in the city) we are forced also to confront the way in which modern capitalist society is evolving as a whole. It makes it impossible for us to treat the problems of small food producers, or of poorly paid workers in rural tourism, as if these were unrelated to the way in which the rest of society's members lead their lives — the sort of food they eat, for example, or the sorts of recreation they expect to enjoy.

The political economy approach emphasises interconnections — between rural and urban, between social and economic, between space and society. Thus it encourages us to locate concerns about rural poverty within a wider debate on what sort of countryside — what sort of rural environment or rural space — we want to achieve in contemporary Ireland: desertification and wilderness in some places, intensive, concentrated exploitation for mass consumption in the rest? or a living, inhabited landscape whose forms reflect the varied ways in which human production interacts with nature? The countryside we get is crucially dependent on the sort of rural society we decide to sustain into the future.

REFERENCES

Bonanno, A. et al., (eds) (1994) *From Columbus to ConAgra — the Globalisation of Agriculture and Food*, University Press of Kansas, Lawrence, KS.

Bord Failte (1994) *Developing Sustainable Tourism — Tourist Development Plan 1994-99*, Bord Fáilte, Dublin.

Brody, H. (1973) *Inishkillane — Change and Decline in the West of Ireland*, Allen Lane, London.

Brundtland Commission (World Commission on Environment and Development) (1987) *Our Common Future*, Oxford University Press, Oxford.

Buck, D., Getz, C. and Guthman, J. (1997) "From farm to table: the organic vegetable commodity chain of Northern California", *Sociologia Ruralis*, 37(1), 3-20.

Callan, T., Nolan, B., Whelan, C.T. and Hannan D.F. with Creighton, S. (1989) *Poverty, Income and Welfare in Ireland*, ESRI Research Series No. 146, The Economic and Social Research Institute, Dublin.

Callan, T., Nolan, B. and Whelan, C.T. (1994) "Who are the poor?", in Nolan, B. and Callan T. (eds) *Poverty and Policy in Ireland*, Gill and MacMillan, Dublin.

Commins, P. (1996) "Agricultural production and the future of small-scale farming", in Curtin, C., Haase, T. and Tovey, H. (eds) *Poverty in Rural Ireland*, Oak Tree Press, Dublin.

Commins, P. and Keane, M.J.(1994) "Developing the Rural Economy — Problems, Programmes and Prospects", in *NESC Report No. 97*, National Economic and Social Council, Dublin.

Curtin, C. and Shields, D.(1988) "The legal process and the control of mining development in the West of Ireland", in Tomlinson, M., Varley, T. and McCullagh, C. (eds) *Whose Law and Order?* Sociological Association of Ireland, Dublin/Belfast.

Curtin, C., Haase, T. and Tovey, H. (eds). (1996) *Poverty in Rural Ireland*, Oak Tree Press, Dublin.

Department of Environment (1997) *Sustainable Development — A Strategy for Ireland*, Stationery Office, Dublin.

Douthwaite, R. (1996) *Short Circuit — Strengthening Local Economies for Security in an Unstable World*, Lilliput Press, Dublin.

Duggan, C. (1994) *Economic Diversity on Smallholdings: a Study of Pluriactivity in West Connemara*, Unpublished Ph.D. thesis, Trinity College Dublin (Department of Sociology).

Dwyer, J. (1991) Structural and evolutionary effects upon conservation policy performance: comparing a UK National and a French Regional Park, *Journal of Rural Studies*, 7(3), 265-75.

Frawley, J. and Commins, P.(1996) *The Changing Structure of Irish Farming*, Rural Economy Research Series No. 1, Teagasc, Dublin.

Goodman, D. (1991) "Some recent tendencies in the industrial reorganisation of the agri-food system", in Friedland, W.H. *et al.* (eds) Towards a New Political Economy of Agriculture, Westview Press, Boulder, Colorado.

Goodman, D., Sorj, B. and Wilkinson, J. (1989) *From Farming to Biotechnology*, Basil Blackwell, Oxford

Heffernan, W. (1984) "Constraints in the US poultry industry", in Schwarzweller H.K. (ed.) *Research in Rural Sociology and Development, Vol. 1 — Focus on Agriculture*, JAI Press, London/New York.

Heffernan, W. and Constance, D.H. (1994) "Transnational corporations and the globalisation of the food system", in Bonanno A. *et al.* (eds) *From Columbus to ConAgra — the Globalisation of Agriculture and Food*, University Press of Kansas, Lawrence, KS

Marsden, T., Murdoch, J., Lowe, P., Munton, R. and Flynn A. (1993) *Constructing the Countryside*, UCL Press, London.

McMichael, P. (1994) "GATT, global regulation and the construction of a new hegemonic order", in Lowe, P., Marsden, T. and Whatmore, S. (eds) *Regulating Agriculture*, David Fulton, London.

Mormont, M. (1990) "Who is Rural? or, How to be Rural? Towards a sociology of the rural", in. Marsden, T. *et al.* (eds) *Rural Restructuring — Global Processes and their Responses*, David Fulton, London.

National Economic and Social Council (1994) *New Approaches to Rural Development*, NESC Report No. 97, NESC, Dublin.

Scott, J. (1995) *Development Dilemmas in the European Community — Rethinking Regional Development*, Open University Press, Buckingham.

Tovey, H. (1991) "Of cabbages and kings": restructuring in the Irish food industry, *Economic and Social Review* 22(4), 333-50.

Tovey, H. (1994) "Rural management, public discourses, and the farmer as environmental actor", in Symes, D. and Jansen A.J. (eds.) *Agricultural Restructuring and Rural Change in Europe*, Wageningen Agricultural University, Wageningen.

Tovey, H. (1996) "Natural resource development and rural poverty", in Curtin, C., Haase, T. and Tovey, H. eds. (1996) *Poverty in Rural Ireland*, Oak Tree Press, Dublin.

6

INTEGRATION AND EXCLUSION IN RURAL IRELAND

James A. Walsh
Department of Geography
National University of Ireland, Maynooth

This chapter addresses three questions: what is meant by rural poverty?, what are the underlying processes? and what solutions can be suggested? The first will be considered very briefly as it has been discussed by contributors elsewhere in this book.

RURAL POVERTY

Approaches to poverty research can be categorised along three axes: absolute versus relative; objective versus subjective, and materialist versus social contextual. Each implies a different perspective which in turn leads to different types of policy prescriptions. The absolute approach is based on the notion that it is possible to define some minimally acceptable standard of living to which all citizens are entitled. This approach focuses the attention on the individual as a separate entity from society and considers that a redistribution of resources from the "haves" to the "have-nots" until the basic needs of everybody have been catered for is an appropriate policy response. The basic needs are also viewed from a very narrow materialistic perspective.

It is more appropriate to think of each individual as part of a large complex social system where the position of the individual is measured against some norms for that society. From this perspective poverty is a relative condition and the focus of attention shifts to the relationship between the individual and society at large.

Research and policy analysis tend to focus on the factors underlying deviations from some measurable objective norms, which are sometimes described as poverty lines.

The so-called "objective" approach has dominated most of the poverty research until very recently. Reliance is placed on statistical indicators that are relatively easily obtained for aggregates of individuals or households. However, quantitative data on their own are an inadequate source for unravelling people's experience of poverty or deprivation or for identifying correctly the categories of persons that are most likely to experience poverty or exclusion — e.g. women, long-term unemployed, youth and early school leavers, elderly without access to private transport, lone parents, people with disabilities, travellers. Recent research in the UK has cast serious doubts on the appropriateness of exclusive reliance on traditional indicator analyses as a means of examining rural poverty and disadvantage — see for example Shucksmith et al. (1996) and Cloke et al. (1997). One of the key conclusions from these research projects is that a rural development strategy based on an area approach may not be particularly helpful on its own if it is not combined with a target group approach. This conclusion is especially relevant at this juncture in the debate on rural policy in Ireland.

The emphasis on tapping into the subjective experience of those who might be deemed to be disadvantaged according to some objective criteria has coincided with a shift of focus from the materialistic concerns (e.g. income, employment, housing, transport) to consideration of a larger agenda that takes account of the social and political processes that lead to the deprivation of rights and exclusion of certain groups from decision making. The latter approach is guided by principles such as enhancement of the self esteem of all individuals by giving greater recognition to diversity and by establishing conditions whereby empowerment actions help to liberate the disadvantaged from the shackles that have become associated with a political culture that has fostered dependency. One of the outcomes of adopting a process, as distinct from a conditions, perspective is that larger issues become part of the analysis: e.g. the concept and process of development; the key processes underlying social, economic, political and cultural transformation; and the capability of existing institutional arrangements to redirect the outcomes from processes that in many cases are driven by forces over which there is relatively little scope for exerting local influence.

UNDERLYING PROCESSES

From the foregoing discussion it is clear that the incidence of rural poverty is influenced by many factors operating at different scales (local, national, supranational, global), and across many sectors, on individuals and groups whose circumstances were to some extent created through the operation of different processes in the past. The processes underlying the contemporary patterns of adjustment in rural areas have been extensively examined by among others Bowler *et al.* (1992), Commins and Keane (1994) and Curtin *et al.*, (1996). For convenience the array of processes can be grouped as follows:

- Economic processes linked to Europeanisation and globalisation
- Social processes affecting demography, and patterns of social integration and interaction
- Political processes reflecting changing ideologies on the role of the state
- Technological processes affecting the nature and organisation of economic activities.

It is only possible to touch very briefly here on some of the main attributes and effects of these processes. For further details see Mernagh and Commins (1997).

Europeanisation and Globalisation

The following are some of the attributes of globalisation and Europeanisation processes:

- Increased emphasis on competition and removal of barriers to market forces
- More attention to efficiency concerns in the design and implementation of public policies
- Increased international and inter-regional competition for mobile investments

- Greater domination of food production and processing industries by fewer large scale units
- Improved accessibility between major centres of population resulting in threats to more remote and sparsely settled areas
- Less local control over factors impacting on development.

Extensive restructuring within agriculture in Ireland has impacted most severely on small farms operated by elderly farmers specialising on cattle and sheep systems, especially in the West and North-West. CAP reform measures have reinforced the tendencies towards larger scale units and increased the reliance of the marginalised on direct subsidies (Commins, 1997). The future viability of many farm households is now very much dependent on household members being able to secure alternative sources of income. The likelihood of being able to do so varies between households due to differences in demographic structure, skills available, and location (Frawley and Commins, 1997).

The manufacturing sector has been characterised by a significant dualism for many years. Apart from the food processing sector, the majority of the most competitive sub-sectors are made up of branch plants of overseas controlled corporations which are increasingly located in, or adjacent to, the larger urban centres. In the era of strong regional policy in the 1970s there was a high level of dispersal of new manufacturing plants which, when coupled with a vibrant agricultural sector in the years immediately following accession to the European Community, led to significant levels of new employment opportunities in rural areas (Walsh and Gillmor, 1993). Since the early 1980s the regional dimension in industrial policy has been given less prominence, reflecting in part the refocusing of strategy towards the achievement of national goals in the European context. The pattern of widespread dispersal of new investments throughout the state has been largely replaced by a more concentrated strategy as there are relatively few locations that can satisfy their soft and hard infrastructural needs (Walsh, 1995; 1997a).

The service sector is the most rapidly expanding employment area. While there has been a general expansion across all regions over recent years, the growth in demand for highly educated graduates is mostly concentrated in the major urban centres, especially Dublin. Tourism has for long been seen as an activity that can make significant contributions to the rural economy and to the maintenance of rural employment levels. However, the evidence in this regard is not entirely reassuring. Most of the recent growth has occurred in those parts of the country that already had a strong tourism base or alternatively in Dublin. There has been little progress in establishing a thriving sector in the weaker regions. Even where tourism has been well established there are reservations over the quality, duration and level of wages associated with much of the associated employment (Hannigan, 1997).

Social Change
The most notable and easily measurable aspect of social change is the demographic adjustment that has occurred especially since the early 1980s (Walsh, 1991; 1996). A very sharp fall in fertility coupled with a very high level of net migration throughout most of the 1980s have brought Irish demography into the greying phase of the demographic transition (McDonagh and Walsh, 1995). Of course rural areas have had disproportionate numbers of elderly people for many decades due to selective out-migration (Horner *et al.*, 1987). Much of the literature on rural poverty draws attention to the plight of the elderly who frequently end up living alone, or in households without any young members, in locations that are distant from most services, and with limited opportunities for interacting with others in the immediate neighbourhood (O'Mahony, 1986). Increasingly, the rural elderly feel threatened and insecure as the traditional more intensive local social networks have broken down. This breakdown can probably be partly associated with the emergence of newer types of younger households with smaller families, high levels of mobility and increasingly dual or multiple occupations. The decline in fertility also has obvious implications for the survival of both primary and second level schools which are often regarded as important focal points for rural communities.

Another feature of social change that has implications for the long term survival of rural populations is the impact of the trend towards certification for entry into the labour force. For well over a decade there has been increased emphasis on the need for educational or training qualifications to gain entry to a highly competitive labour market. In response to this trend more and more students are opting to remain in the educational system for longer periods and seeking to acquire third level qualifications. The relevance of this trend for the rural population is that as education levels increase so also do the levels of inter-regional and international migration, with the decision to attend college often being the first link in a migration chain (Hannan, 1992; McHugh and Walsh, 1995).

There are indications of in-migration into some rural areas. While some may be categorised as return migration (about which very little systematic knowledge has been accumulated) there is also in some scenically attractive rural areas a high level of immigration of persons without any previous Irish connections. One of the concerns that has been expressed about such inward migration is that it can contribute to local inflation of housing and land values, making it more difficult for local residents to acquire land to extend their holdings or to acquire housing. The housing issue is compounded by the proliferation of holiday homes.

The social issues that are noted here are well documented. However, there is a need for policy responses that address the implications of the changes in demography, and the implications of the economic adjustments for future settlement patterns.

Changing Ideologies on the Role of the State

The processes of economic restructuring that have been underway since the early 1970s have been accompanied by a revision of the dominant ideologies concerning the role of the state in western democracies. The transition to post-Fordism in an era of unprecedented levels of unemployment has contributed to the emergence of a neo-liberal perspective on the role of the state in relation to the economy. This has been accompanied by greater emphasis on privatisation and deregulation. More traditional concerns with intra-state regional policies have been largely abandoned as the effects of restructuring and the pressures related to demographic growth became more pronounced in large urban areas. The shift in focus to seeking sectoral solutions has given greater prominence to efficiency criteria in the allocation of public resources, which has resulted in the "rationalisation" of the delivery of many services — e.g. health, education, security, postal services. These adjustments have impacted severely on many rural residents, particularly the elderly, persons with disabilities, and those without access to private transport facilities.

At the same time, the scope for action by the state is being limited by pressures from both above and below. Much of the policy direction and many of the programmes that are currently being implemented are linked to EU policies and programmes. At the other end of the political/administrative system there has emerged a strong sense of localism. This has come about partly in response to the mega tendencies in the economic, social and political spheres which have left many local communities more vulnerable and threatened. In response there have been intensive efforts to establish, or in some cases re-establish, local identities that can provide an effective means of responding to niche markets in an increasingly differentiated consumer market. The emergence of localism is also driven by a recognition of the need to give greater effect to the principle of subsidiarity by establishing local structures that facilitate participation by a wide range of interests. There has been much experimentation with local development initiatives over recent years (Walsh, 1997b) which has been favourably reviewed by the OECD (1996), though, perhaps not sufficiently critically (Walsh, 1996b). However, it is important to note here that only one of the local development initiatives has a specific remit to target those who are socially excluded or in danger of becoming so (Government of Ireland, 1995). It is a mistake to assume that because programmes are administered locally that the resources will be targeted to those in greatest need — often in reality it is those who are already well resourced that have the capacity to link into new sources of public funds.

A final point of concern that arises out of the emergence of localism is the balance in the sharing of responsibility between the state and local groups for addressing major problems such as rural poverty. There is a need for a greater sense of realism in regard to what can be achieved through local partnerships and also in regard to how their efforts might be measured in evaluation studies (Walsh, 1996c). There are also important issues concerning the relationship between local partnerships and local government structures (Coyle, 1997, Department of the Environment, 1996).

Technological Change

Technological change underpins many of the adjustments that are occurring in other spheres of activity. All new technologies bring opportunities and threats. In the productivist phase of agricultural development new technology tended to displace labour. More importantly, the processes involved in technology adoption have tended to favour large-scale modernising farms and simultaneously render many small farms less economically viable. The geography of technical change has been mediated by many factors resulting in significant long-term contrasts between the East and South-East versus the West and North-West (Walsh, 1992).

The potential impacts on rural society of the most recent technological revolution remain unclear. While undoubtedly many opportunities will arise in the production of hardware and software, and in a diverse range of applications there are some grounds for concern over the locational implications of the revolution in information and telecommunications technology (Grimes, 1997). While some may argue that geography does not matter anymore in the era of internet, this is unlikely to be the situation. IT is an enabling technology which is more likely to reinforce existing spatial divisions of labour and patterns of rural social exclusion. A key issue that must be considered is the wider economic and social milieu requirements for the most efficient use of the new technologies. In the context of a discussion on rural poverty, one of the challenges is to identify ways in which disadvantaged or excluded persons can be equipped to benefit from the opportunities that may arise.

A FRAMEWORK FOR TACKLING RURAL POVERTY AND EXCLUSION

Any attempt to address the problems of poverty and exclusion in rural areas must take place within a framework for rural development that explicitly recognises the need for an inclusive strategy. A critical area that must be addressed in the design and implementation of the strategy is the effectiveness of the current array of institutions that are responsible for the formulation and delivery of policies and programmes.

The term *institution* is used here to refer to any collective body — whether it be a government department, state agency, local partnership company, local authority, private business sector representative association, or voluntary organisation/grouping to represent specific community interests — that has a role in formulating and delivering policies and programmes to facilitate a socially inclusive and sustainable approach to rural development. Institutional arrangements are taken to be concerned with the organisational structure of the relevant bodies, and the mechanisms for establishing co-ordination linkages.

Outline of Present Institutional Arrangements for Rural Development

At present there are a very large number of institutions involved in performing many diverse roles in support of rural development. These include:

- The European Commission which has a role in promoting EU-wide policies that directly or indirectly impact on rural areas. The Commission also provides much of the funding for programmes that impact on rural areas in Ireland.

- Government Departments which are responsible for policy formulation, and in some cases delivery of programmes.

- State agencies with specific sectoral responsibilities.

- Local Authorities with responsibility for local infrastructure, local physical planning and provision of some public services.

- Regional Authorities with responsibility for inter-county co-ordination of the delivery of public services.

- Regional Agencies (Shannon Development and Udaras na Gaeltachta) with responsibility for supporting selected aspects of economic development in their areas.

- The Western Partnership Development Board with responsibility for co-ordinating the delivery of programmes in a number of western counties.

- Local partnerships with responsibility for the implementation of local development programmes. These include County Enterprise Boards, Area-Based Partnerships to implement integrated programmes for the development of Disadvantaged and other areas, and LEADER groups.

- County strategy groups to secure cohesion at county level of the various local development initiatives being undertaken by the State and voluntary sectors.

Weaknesses in Current Arrangements

The present institutional system suffers from several well documented weaknesses that significantly curtail the potential for more inclusive approaches to rural development (Walsh, 1996d). These include the following:

1. Top-down structure: most of the current decision-making takes place within a top-down framework.

2. Sectoral, compartmentalised policies and programmes, resulting in an absence of adequate prioritising of objectives, and conflicts with regard to resource allocation and utilisation.

3. Economic development focus: policies, plans and programmes have traditionally had a primarily economic focus with limited direct targeting towards the needs of the disadvantaged.

4. Inadequate co-ordination: there are inadequate mechanisms for both horizontal and vertical co-ordination of the different agencies involved.

5. Representation not sufficiently inclusive: the groups in society which are typically affected by social exclusion tend not to be represented, or at best are under-represented on the principal decision-making bodies. There is also a democratic deficit in the composition of most partnership boards.

Proposals for an Inclusive Institutional System

There is a need for a strategically organised institutional response to a number of instances of market failure and social exclusion which can lead to sub-optimal levels of economic activity and social participation. Many of the instances of market failure that occur have regional or local dimensions. A support framework is required that can respond in a satisfactory manner to criteria such as equality, inclusiveness, partnership, effectiveness, accountability and subsidiarity. The institutional framework required to provide a more effective system of supports for rural development has to address the weaknesses already identified. Much attention has been directed to these issues over recent years — see, for example, the National Economic and Social Council Report on *New Approaches to Rural Development* (1994), the report of the National Economic and Social Forum on *Rural Renewal: Combating Social Exclusion* (1997), *the Report of the Rural Development Policy Advisory Group* (1997), and the National Anti-Poverty Strategy *Sharing in Progress* (1997).

There are a number of actions that can be undertaken to improve the current system by focusing on the following issues.

1. Adoption of a Multi-Dimensional Concept of Development by Decision-makers at All Levels

This will require a replacement of the view that development is an incremental process primarily driven by economic considerations, by a perspective that regards it as a process of structural change which empowers individuals and groups to pursue agreed socio-economic, cultural and environmental objectives in a sustainable manner — this perspective on rural development has been promoted by both the OECD (1990) and more recently the European Commission in the Cork Declaration on *A Living Countryside* (1996). In this approach the primacy of short term economic considerations, supported by a mix of public policies that may contribute to further exclusion and dependency, is replaced by concepts such as participation, integration and long term sustainability. In order to promote the new concept of development, a National Level Partnership for Rural Development should be established with responsibility for ensuring that the policies and strategies of all government departments and public agencies conform to principles of multi-dimensional, inclusive and sustainable development opportunities for all rural dwellers.

2. Equality Proofing

Over recent years there have been a number of reports which have advanced arguments in support of the principle that all government policies and programmes should be subjected to equality proofing in their design and implementation (NESF, 1996). The application of this principle to the design and quality of policies and programmes, and also to institutional representation, would assist in preventing the occurrence of rural social exclusion.

3. Inclusive Framework for Integrated Spatial Planning

There is a need for an integrated spatial planning framework extending from national through regional to local level. The conventional approach to development planning is for the most part organised from the centre along sectoral lines and targeted towards crudely defined rural and urban areas, with very limited scope for local involvement. This approach needs to be refined to take account of different types of rural areas, the linkages between rural and urban areas, and the structure of the settlement pattern. Otherwise, there is a risk that excessive responsibility will be placed on inadequately resourced local partnerships to resolve problems that emanate from outside their areas. In keeping with the NESC recommendations, the goals of rural development must be addressed within the context of balanced regional and local development and national settlement patterns.

4. Adoption of Partnership as an Organisational Model

For a variety of reasons, recent experience in Ireland and many other OECD countries suggests that partnerships of local groups, private interests, government and statutory agencies provide innovative organisational models for sensitive, flexible and effective implementation of rural policies and programmes. The partnership model provides a means of bringing together the various stake-holders in the rural development process. The partnering process represents a fundamental shift from the traditional hierarchical mode of governance to a negotiated approach. In order to work effectively and in an innovative manner, partnerships must be guided by a commitment to a shared vision and a willingness on the part of all partners to re-evaluate their own agendas in the light of the shared vision. To date, the partnership approach has been largely confined to local programmes and to the negotiation of national level development priorities. The experience at regional level in relation to roles of Regional Authorities has been less than satisfactory (Fitzpatrick Associates, 1997).

There are several difficulties that need to be addressed before the full potential of the partnership approach can be realised. These have been highlighted in a number of recent evaluation studies and comprehensively addressed in a report from the National Economic and Social Forum (1997b). Successful implementation of the partnership approach is very dependent on building trust between partners; willing and proactive participation by all members; a matching of local partnership arrangements by similar ones at central (state) level; and greater attention to the range of interests that are represented either directly or indirectly on boards. Partnership board members require technical support in areas such as negotiation and communication skills, in order to facilitate a high level of sharing of experience. The principles that underpin the formation and effective functioning of partnerships should be adopted by institutions at all levels.

5. Better Targeting of Support Programmes

There is need to ensure that the area-based strategies implemented by local partnerships are better targeted so that specific actions are included to proactively facilitate a process of empowerment of those who have been excluded. At present the programme administered by ADM for Integrated Development of Designated and Other Areas of Disadvantage is the only one with a specific remit to target the socially excluded. It supports targeted education and training measures for both early school leavers and adults, a locally operated contact service for the unemployed, and a range of actions to assist community development.

One of the weaknesses of the current array of local development initiatives is that it is highly compartmentalised on a departmental basis, uncoordinated in terms of the spatial units adopted by different agencies, incomplete in terms of areal coverage in that only the most disadvantaged rural areas are eligible to participate in the programme administered by ADM, and likely to encourage inter-agency competition rather than co-operation in the pursuit of scarce resources. In practice the measures to address social exclusion should not be confined solely to the programme administered by ADM, rather they should be available to communities throughout the State. The problems that arise in this regard are due in part to the adoption of an area-based approach to the delivery of the programme where disadvantaged areas are identified on the basis of the severity of deprivation as measured by a multivariate statistical index (see Haase in this volume). Better targeting is required to ensure that individuals or groups are not further disadvantaged by virtue of their location within socially mixed spatial units such as District Electoral Divisions that are used as the basic geographical units in the analyses leading to the calculation of deprivation indices.

6. Procedures for Horizontal and Vertical Co-ordination
One of the most significant weaknesses in the current institutional system is the limited capacity for horizontal and vertical co-ordination. Mechanisms for horizontal co-ordination at county level have been put in place through the County Strategy Groups. These are likely to be revised as some form of Community and Enterprise Groups are established (Department of the Environment, 1996). At regional level the Regional Authorities have a responsibility for co-ordinating the provision of public services. At national level the National Anti-Poverty Strategy (NAPS) proposed that a team should be established as a Strategic Management Initiative within the Department of Social Welfare to implement the strategy. At the same time it was proposed that a cabinet sub-committee should be established to deal specifically with issues of poverty and social exclusion. Furthermore, responsibility for overseeing an evaluation of the NAPS process has been assigned to the Combat Poverty Agency. All of the initiatives described here are of very recent origin, and consequently are likely to require some time before the synergies that might be expected from co-ordination will be become evident. The experience to date suggests that a considerable effort will be required to ensure that these initiatives will result in some real changes that will directly address the underlying causes of poverty and exclusion in rural areas.

Vertical co-ordination refers to co-ordination of activities undertaken by agencies operating at different levels. The issues involved here are probably greater than those at the horizontal levels. At present, the available mechanisms include: intra-agency structures as many of the agencies responsible for delivery of the mainstream programmes are structured on a regional basis with additional local offices in county towns; and State-local structures to link the agencies responsible for delivery of the local development programmes with their parent departments or with an Intermediary Body appointed by the Department (e.g. ADM). The effectiveness of both types of structures is influenced by the extent of subsidiarity and the representativeness of the structures that are being co-ordinated.

There are major weaknesses in the vertical chain at the regional level. The Regional Authorities are not effectively linked to any co-ordinating group at the centre. Furthermore, the State does not have a clear policy in relation to regional development. At the lower end of the scale there is no mechanism for linking local initiatives upwards to regional strategies. There are also situations where local is not sufficiently close to the target groups. In many parts of the state, local development is organised at county level without adequate sub-county structures to facilitate effective participation of a wide range of interests. The issues of co-ordination and integration must be tackled through strategic alliances between the different agencies responsible for rural development polices and programmes. Horizontal and vertical co-ordination should be an explicit responsibility of partnerships at different levels.

7. Representation

One of the key issues in recent reviews of local development in Ireland is the question of a democratic deficit in many local partnerships. This is also an issue that was considered by the Devolution Commission (1997). A resolution of the issue probably lies in including a small number of elected local councillors on the boards of all local partnerships. Attention must also be directed towards ensuring that other board members of partnership companies are representative of well defined interests and that the partnership structures provide opportunities for the interests of the socially excluded to be represented.

CONCLUSION

This chapter has argued that a multidimensional perspective is required to tackle the processes that contribute to the plight of disadvantaged and excluded persons in rural areas. The root causes of many of the problems lie in processes operating in the economic, social and political spheres which at times are orchestrated by interests that are organised on a supra national basis in pursuit of objectives that frequently have little to do with poverty. In the final part of the chapter attention was focused on the limitations of the current institutional framework which was considered to be elaborate, inefficient and for the most part incapable of addressing the issues related to social exclusion in rural development. Any serious attempt to promote a more socially inclusive approach to rural development must tackle in an integrated way the range of issues identified above. While there have been some modest signs of innovation in recent times much more is necessary so that a more inclusive, participative and sustainable model can replace the conventional approaches that tend to be divisive, paternalistic and conducive to a widespread sense of dependency.

REFERENCES

Bowler, I. R., Bryant, C.R. and Nellis, M.D. eds. (1992) *Contemporary Rural Systems in Transition. Volume 1: Agriculture and Environment*, CAB International, Wallingford.

Cloke, P., Goodwin, M. and Milbourne, P. (1997) *Rural Wales — Community and Marginalization*, University of Wales Press, Cardiff.

Commins, P. and Keane, M.J. (1994) Developing the Rural Economy: Problems, Programmes, Prospects, in National Economic and Social Council, *New Approaches to Rural Development*, Part 11, NESC, Dublin.

Commins, P. (1997) Agricultural Production and the Future of Small-Scale Farming, in Curtin, C. *et al.* (eds.), *Poverty in Rural Ireland*, Oak Tree Press, Dublin.

Commission of the European Communities (1996) *Cork Declaration — A Living Countryside*, Brussels.

Coyle, C. (1997) Local and Regional Administrative Structures and Rural Poverty, in Curtin, C. *et al.* (eds.), *Poverty in Rural Ireland*, Oak Tree Press, Dublin.

Curtin, C., Haase, T. and Tovey, H. (eds.) (1996) *Poverty in Rural Ireland*, Oak Tree Press, Dublin.

Department of the Environment (1996) *Better Local Government: A Programme for Change*, Government Publications Office, Dublin.

Devolution Commission (1997) *Second Report*, Government Publications Office, Dublin.

Fitzpatrick Associates (1997) *Mid-Term Evaluation of Sub-Regional Impact of Community Support Framework 1994-1999*, Fitzpatrick Associates, Dublin.

Frawley, J. and Commins, P. (1997) *The Changing Structure of Irish Farming: Trends and Prospects*, Teagasc, Dublin.

Government of Ireland (1995) *Operational Programme for Local Urban and Rural Development, 1994-1999*, Government Publications Office, Dublin.

Government of Ireland (1997) *Sharing in Progress: National Anti-Poverty Strategy*, Government Publications Office, Dublin.

Grimes, S. (1997) "The implications of information technologies for regional development in Ireland", in McCafferty, D. and Walsh, J. (eds.) *Competitiveness, Innovation and Regional Development in Ireland*, Regional Studies Association (Irish Branch), Dublin, 203-212.

Hannan, D. (1992) "Education, employment and local economic development", in Davis, J.P. (ed.) *Education, Training and Local Economic Development*, Regional Studies Association (Irish Branch), Dublin, 25-33.

Hannigan, K. (1997) "Tourism policy and regional development in Ireland", in McCafferty, D. and Walsh, J. (eds.) *Competitiveness, Innovation and Regional Development in Ireland*, Regional Studies Association (Irish Branch), Dublin, 171-92.

Horner, A., Walsh, J.A. and Harrington, V. (1987) *Population in Ireland: A Census Atlas*, Department of Geography, University College Dublin, Dublin.

McDonagh, P. and Walsh, J.A. (1995) "A new demographic revolution — changing age structures in the European Union", *Geographical Viewpoint*, 23, 111-24.

McHugh, C. and Walsh, J.A. (1995) "The Irish school-leaver trail: links between education and migration", *Geographical Viewpoint*, 23, 88-103.

Mernagh, M. and Commins, P. (1997) *In from the Margins*, SICCDA, Dublin.

National Economic and Social Council (1994) *New Approaches to Rural Development*, Report No. 97, NESF, Dublin.

National Economic and Social Forum (1996) *Equality Proofing Issues*, Report No. 10, Government Publications Office, Dublin.

National Economic and Social Forum (1997a) *Rural Renewal — Combating Social Exclusion*, Report No. 12, Government Publications Office, Dublin.

National Economic and Social Forum (1997b) *A Framework for Partnership — Enriching Strategic Consensus Through Participation*, Report No. 16, Government Publications Office, Dublin.

OECD (1990) *Partnerships for Rural Development*, OECD, Paris.

OECD (1996) *Local Partnerships and Social Innovation in Ireland*, OECD, Paris.

O'Mahony, A. (1986) *The Elderly in the Community: Transport and Access to Services in Rural Areas*, National Council for the Aged, Dublin.

Shucksmith, M., Chapman, P. and Clark, G.M. (1996) *Rural Scotland Today — the best of both worlds?* Avebury, Aldershot.

Walsh, J.A. (1991) "The turn-around of the turn-around in the population of the Republic of Ireland", *Irish Geography*, 24(2), 116-24.

Walsh, J.A. (1992) "Adoption and diffusion processes in the mechanisation of Irish agriculture", *Irish Geography*, 25(1), 35-53.

Walsh, J.A. (1995) *Regions in Ireland: A Statistical Profile*, Regional Studies Association (Irish Branch), Dublin.

Walsh, J.A (1996a) "Population change in the 1990s: preliminary evidence from the 1996 Census of Population in the Republic of Ireland", *Geographical Viewpoint*, 24, 3-12.

Walsh, J.A. (1996b) "Review highlights need for partnership reforms", *Poverty Today*, 33, 12-13.

Walsh, J.A. (1996c) "Local development theory and practice: recent experience in Ireland", in Alden, J. and Boland, P. (eds.) *Regional Development Strategies: A European Perspective*, Jessica Kingsley Publishers, London.

Walsh, J.A (1996d) "Institutional arrangements for regional development", in *Shannon Development, Regional Policy*, Shannon.

Walsh, J.A. (1997a) "Regional development challenges", in McCafferty, D. and Walsh, J. (eds.) *Competitiveness, Innovation and Regional Development in Ireland*, Regional Studies Association (Irish Branch), Dublin.

Walsh, J.A. (1997b) "Development from below: an assessment of recent experience in rural Ireland", in Byron, R., Walsh, J. and Breathnach, P. (eds.) *Sustainable Development on the North Atlantic Margin*, Ashgate, Aldershot.

Walsh, J.A. and Gillmor. D. (1993) "Rural Ireland and the Common Agricultural Policy", in King, R. (ed.) *Ireland, Europe and the Single Market*, Geographical Society of Ireland, Dublin.

7

POVERTY AND ACCESSIBILITY TO SERVICES IN THE RURAL WEST OF IRELAND

Mary Cawley
Department of Geography, National University of Ireland, Galway

A RATIONALE FOR STUDYING ACCESSIBILITY TO SERVICES

Populations depend for their social and economic well-being on a wide range of services which are provided by public and private agencies, by voluntary organisations and by individual entrepreneurs. These include basic education, health, transport, housing, recreational, religious, and law enforcement provision but the list is virtually endless. Issues of economic efficiency underpin the structures and mechanisms through which services are provided. Some cognisance is usually taken of equity considerations also, most notably in the case of public welfare services. In response to restrictions on public expenditure, the quest for economies of scale in both public and private service delivery, and increasing privatisation of state utilities, restructuring of service provision has taken place in many countries in recent years (OECD, 1991). A reduction in the number of service delivery points has been one feature of restructuring which is associated with rural areas in particular. Such reductions have direct implications for the accessibility of services to low density rural populations: the average travel time and costs associated with service use inevitably increase for substantial numbers of consumers in the wake of closures. Certain sub-groups within these populations are affected disproportionately. These include the elderly who are dependent on state pensions, who frequently lack access to private transport and who may be unable, for physical reasons, to use a public transport service if such is available; persons on low incomes in general; and women with young children who do not have access to private transport during the working day when a family-owned vehicle is in use for work-related purposes (Fyfe, 1994). Studies in other countries document the lowered access to services now being experienced by rural populations in general and their negative implications for the less-well-off in particular (see, for example, Batten and Wiberg, 1988 for Sweden).

Issues of societal and individual welfare require that more attention should be given to accessibility to services. Negative impacts may result from poor accessibility for health and for educational attainment as well as for the general quality of life experienced in rural areas. Long-term implications relate to the lowered ability of such areas to retain their populations and to maintain viable economic and social systems. The lower densities of population in rural areas means that fewer provision points are available than in urban areas. Travel to use a particular service almost invariably involves a greater expenditure of time and money for a user than in an urban location. The time and cost incurred by a service provider is also greater in a rural area because distance reduces the number of dispersed consumers who can be visited in any given time period. The focus of this chapter is on the accessibility of basic services which are either delivered at fixed points, which requires that consumers travel to avail of them, or are delivered to the consumer in his or her home with the deliverer undertaking the travel. Housing, which was included in an earlier survey on this theme, is omitted here because it differs in its method of delivery (Cawley, 1986). Housing is also a specific focus of another chapter in this book.

This chapter proceeds by reviewing (a) broad trends in the delivery of a number of basic services in the Republic of Ireland in recent years, (b) the problems of accessibility experienced by low-income rural populations as revealed in empirical research, and (c) remedial strategies that are currently in place or planned to offset low levels of accessibility. Broad issues that appear to merit attention in the context of improving accessibility are identified.

CHANGING TRENDS IN SERVICE DELIVERY

For some three decades a gradual contraction has been taking place in the network of service provision points in rural Ireland. This trend is related in part to declining population thresholds as a result of outmigration. Rural shops were among the first services to be affected by population decline, as were rail services as use declined associated with rising levels of private car ownership and rising use of road freight haulage (McKinsey International Inc., 1979). Falling birth rates reduced the number of children entering primary schools and served to lower rolls in some schools below critical thresholds required for economic efficiency by the early 1960s. Difficulties were experienced in attracting young teachers and doctors to primary schools and dispensaries, respectively, in remote locations from the 1960s on because of the weak professional support frameworks that were available and the associated limitations on opportunities for career advancement. Thus, transport and fixed-point services were being lost through a process of gradual attrition during the 1950s and the early 1960s.

Planned closure of services took place during the second half of the 1960s and the 1970s. During this latter period, a policy of closing small rural dispensaries and one- and two-teacher schools was pursued in a dual quest for increased economic efficiency and enhanced levels of service provision (Curry, 1993). "Professionalism" became more marked in the health sector and involved the appointment of specially trained ancillary nursing and health care personnel in health centres served by a number of general practitioners at community level (McCashin, 1982). Economies of scale made it possible to provide higher levels of diagnostic equipment than had been possible in individual rural dispensaries. Medical card holders were also provided (in theory at least) with a choice of physician from 1972 on, an option which they did not have under the dispensary system. Whilst the level of medical services provided increased, accessibility to the services declined for many rural dwellers because of the location of the health centres in small towns and villages.

Centralisation took place in the primary school sector also. Some 540 one-teacher and 1,200 two-teacher schools were closed between 1965 and 1977 through amalgamation with other schools (Curry, 1980). Persuasive arguments in favour of school closure included the provision of improved school buildings and educational aids, contact for the pupil with a larger number of teachers over the school cycle and a curriculum which was to be extended to include art, physical education and environmental education. A free transport system was introduced for children who lived more than two miles from a primary school although this was replaced by means-tested travel in 1983 because of the increasing cost of school transport. This is now subject to withdrawal when the number of students travelling from a particular area falls below five (Western Development Partnership Board, 1996). School closure as an active policy was abandoned in 1977, but the recently published White Paper on Education, *Charting our Education Future* (Department of Education, 1995), refers to a need for further rationalisation because of the falling intake of pupils.

Transport provision assumes particular importance because of its influence on the accessibility of other services. Rail networks serve to link major regional cities with Dublin and in the process provide a link between the towns located on the line which have retained scheduled stops. Provincial bus services consist of two elements, the Expressway service which links major towns with each other and complements the rail network, particularly in a north-south direction, and the rural request-stop service which serves smaller towns and villages. This latter service has contracted over time. Route densities are low and services are infrequent. To compensate for inadequacies in public transport, private minibus services have developed in some rural areas during the last decade and provide transport for a range of activities, including the collection of pensions, attendance at church on a Sunday, and bingo playing. Official government policy has tended to discourage the development of rural transport services, through the refusal of licenses, except in areas where the proposed service meets a need not met by an existing state service (Barrett, 1991). Such restrictions have been avoided in the past by the formation of "travel clubs" which are not subject to special licensing. A concerted rural transport policy as such is absent, but the Green Paper on *Transport Policy* (Department of Communications, 1985) recognised that postbuses, taxis/hackney cars and the use of school buses outside school hours could play a role in meeting transport needs in rural areas. The populations of the offshore islands are doubly disadvantaged in relation to public transport in that they lack an internal transport system and, until very recently, the smaller islands lacked scheduled services (Royle and Scott, 1996). The inadequacies of public transport services in rural areas impinge in particular on the relatively high proportions of households without access to private transport (Byrne, 1991). Even where private transport is present, maintenance costs on vehicles are high because of the relative neglect of second and third class roads during the 1980s when emphasis was placed on improving the National Primary road network to assist economic competitiveness (Barrett, 1991).

The contraction of basic services has been taking place over prolonged periods of time, although the negative effects have reached critical levels in recent years. Closure of delivery points and a reduction in provision became features of several other services in rural Ireland during the 1980s and the first half of the 1990s. Many garda stations in small villages experienced a reduction in personnel during the 1980s and some now have limited opening hours. Car patrols are undertaken from nearby towns and limited resources prevent the dense network of minor roads from being patrolled on a regular basis. Formal policing is supplemented by Neighbourhood Watch schemes co-ordinated by the gardaí. Rural post offices are also at risk from cutbacks in government expenditure. As part of a drive to reduce its debt, An Post mooted a proposal in 1991 to introduce a phased closure of some 550 sub-post offices located in urban and rural areas (*The Irish Times*, February 9, 1991). Many rural residents now have their mail delivered to mailboxes placed on roadsides rather than to their doors. The loss of even sporadic contact with the postman increases the isolation of elderly people. Telephone ownership which would serve to facilitate social contact more generally is notoriously low among the rural elderly. Remote rural areas with low population densities have also, in general, failed to benefit from the development of new services, notably in the areas of recreation, childcare and day-care for the elderly, to the extent that urban areas have done.

ACCESSIBILITY PROBLEMS EXPERIENCED

Formal identification of accessibility to services as a factor that influences the quality of life experienced has, until recently, received relatively little attention from researchers in the Republic of Ireland by comparison with other countries (Storey, 1994). Problem definition is emerging from academic studies which may not necessarily have a policy objective, but the results of which have policy implications, and from policy-related research undertaken or sponsored by a range of state, private and voluntary bodies. In addition, the emergence of lobby groups such as CORPO (Conserve our Post Offices) and CRA (Cavan Road Action group) has highlighted accessibility problems. Some of the difficulties that have been documented are discussed in the context of the key services reviewed above: health, education, transport, and a range of more general public services.

Health Care

The health sector incorporates a wide range of services relating to aspects of health promotion, domiciliary care, residential care, general practitioner (GP) services, hospital outpatient and inpatient care. There is some evidence available in each case which suggests that less than optimal levels of accessibility are experienced by the rural poor and that additional attention needs to be given to designing mechanisms to ensure more effective delivery of, and accessibility to, services. Screening to detect early stages of pathological conditions is recognised as promoting personal well-being and as reducing public health care costs in the longer term. Screening, which may require attendance at a hospital clinic currently, is of particular importance in the case of certain types of cancer. Recent research by Sixsmith (1995) on attitudes of women to mammography for breast cancer in the west of Ireland revealed that rural medical card holders favoured the introduction of a mobile screening unit and the universal availability of screening. Distance, as such, was not mentioned as a factor that would prevent attendance at a clinic by these women but the time and cost involved in overcoming distance were underlying considerations for them.

The domiciliary care provided by the Health Boards includes public health nursing care and home help services. Domiciliary visits by GPs may be considered as a component also. In all instances the care provider travels to the home of the recipient of care. The distribution of the patients influences the number of visits that can be made and the amount of time that can be allocated to each patient in any time period. Certain features of the distribution of population in the Western Health Board area have negative implications in this regard. Population density is low, there are substantial numbers of elderly people living alone in remote areas and the number of public health nurses per 1,000 population is lower than in the Eastern Health Board area which has a much denser distribution of population (Department of Health, 1995). Compensating for the problems of overcoming distance receives attention from the Western Heath Board in its strategic planning (Western Health Board, 1995). The need for institutional care for the elderly also tends to be high in remote rural areas where family support systems have been depleted through emigration (O'Connor et al., 1988). Obtaining a place in a nursing home, however, almost invariably involves a move from his or her home environment and neighbours for the elderly person which may prove stressful.

The absence of a local GP's surgery was identified as entailing additional travel for rural residents in some parts of north-west Connemara, in research conducted by Forum (Byrne, 1991). Postponing consulting a GP in the early stages of an illness may have longer term effects which require admission to a hospital. Similarly, failure to receive dental care at an early stage can result in a need for tooth extraction. In the case of outpatient care, research conducted on attendance at outpatient hospital clinics at University College Hospital Galway reveal some of the main problems associated with attendance for rural residents (Cawley and Stevens, 1987). Long distances were travelled by many patients, substantial amounts of time were spent travelling and high costs were incurred. Patients who were dependent on taxis or came from off-shore islands, when an overnight stay was necessary, incurred particularly high costs: some patients were eligible to have these costs reimbursed but delays in receiving reimbursements were reported. Persons on low incomes and the elderly reported particular difficulties in accessing hospital clinics and were often dependent on others for lifts. In some instances, a patient travelled with a family member who was coming to work in Galway and waited to return with them in the evening. Mothers of young children who encountered difficulties in making childcare arrangements during their absence from home had some of the highest non-attendance rates. In the case of elderly patients, it emerged that transport costs were often borne by family members who may have taken time off work to ensure that a relative kept a hospital appointment. Whilst average distances travelled and costs were substantial, *per se* they did not explain the missing of appointments. The highest rates of non-attendance were registered among patients who had been attending at infrequent intervals over long periods of time and may have mislaid an appointment card or failed to appreciate fully the rationale for long term monitoring of their medical condition. This finding suggests the need for increased attention to be given to the mechanisms of service delivery as well as to their use when solutions to poor accessibility or under-use are being investigated. Distance from hospitals assumes even greater proportions when emergency attention is needed in areas with relatively low availability of ambulances. The provision of an emergency helicopter service is a substantial improvement in increasing accessibility for the residents of the off-shore islands who in the past would have had to travel by trawler and ambulance to reach a hospital.

Education
Few rural areas contain a second level school so that travel for post-primary education is a given feature of rural life (Storey, 1994). Apart from transport costs, which are means-related and therefore should not apply for pupils from low-income households, the inconvenience and loss of time spent waiting for buses along roadsides and travelling are accepted as the lot of the rural student. Attendance at boarding school is necessary in the case of children from most of the off-shore islands. Funding is available in the form of the Remote Area Boarding Grant but the threshold income level for eligibility is felt to be too low (Interdepartmental Co-ordinating Committee on Island Development, 1996). Low density populations, low income and poor access to transport inhibit the capacity of adults over extensive areas of the countryside to benefit from the opportunities for continuing education that are

available to their contemporaries in the larger towns. Pre-school provision is also rare in remote rural areas and may have implications for the later educational progress and employment prospects of the child (Breen, 1991).

Increasing proportions of primary school students are also travelling daily to schools in towns and villages. Some students from off-shore islands make a daily return journey by boat whilst others board on the mainland during the week and return home at weekends (Cork County Council and FÁS, 1994). The counties of western Ireland in general experienced some of the highest rates of school closure and amalgamation during the 1970s and they are also at greatest risk from any renewed closure policy because of the presence of smaller schools and falling population. The views of children, parents and teachers affected by school closure through amalgamation in west County Clare during the late 1960s and the 1970s, as reported by McMahon (1989), document some of the impacts. A lengthening of the period of time spent away from home was reported, arising from travelling longer distances to school. This impinged in particular on pupils in the infant classes. Lack of opportunities for observing wildlife and a loss of casual encounters with friends and neighbours on the journey to and from school were referred to. Contacts with a wider range of teachers and friends were viewed as benefits of the larger school. Few perceived that there was any substantial difference in the provision of educational facilities between the old and the new school, although this was an argument forwarded in favour of amalgamation. Parents were, in general, less positive about the effects of school amalgamation than were their children. They viewed the closure of the local primary school as having had negative effects for social cohesion in removing a point for local contact. Reservations were also expressed about negative social, educational and emotive effects associated with travelling by school bus. Teachers reported lower levels of involvement in community activities in the area where closure occurred after they moved to a new school. They, like the students, reported that the provision of teaching aids was not appreciably superior in the larger schools.

Whilst small rural schools are felt to have social and indeed educational benefits in introducing the student to formal education in a familiar environment, some of these schools lack essential services such as those of a remedial teacher (Irish National Teachers' Organisation, 1995). A recent study relating to southwest County Clare points to links between poverty and educational disadvantage (*The Irish Times*, January 9th 1995). In this particular case there are high levels of unemployment among parents who are therefore unable to contribute to fundraising activities for the schools. Special provision for such schools is necessary.

Transport

Inadequacies in public transport provision, relatively low levels of ownership or poor access to private transport, and the costs of vehicle repair in areas where road surfaces have deteriorated through lack of adequate maintenance, emerge in many studies as impinging on the levels of accessibility to services experienced by rural dwellers. Overcoming the obstacle of distance involves additional inconvenience, time and costs for individuals and families in rural areas. Private taxis or hackney cars are engaged by non-vehicle owners who can afford this option. Those who lack adequate disposable income often depend on neighbours for lifts to keep hospital appointments (Cawley and Stevens, 1987). Even when transport costs may be recouped, some people are reluctant to seek subsidies. The inadequacies of rural public transport services impinge severely on the less-well-off who do not have private motor transport and on those who lack access to a private vehicle during the working day. Holders of free travel passes are frequently unable to avail of their entitlement and may pay for private services instead. Even when a public transport service is available some elderly people are unable to use it because of arthritic problems which prevent them climbing the steps to enter a conventional bus (O'Mahony, 1986). Research among island communities reveals difficulties associated with keeping appointments on the mainland, particularly if a regular transport service is not available or if the service is disrupted due to poor weather conditions (Royle and Scott, 1996). Differences also exist between islands which are related to varying levels of subsidy being available in Gaeltacht and non-Gaeltacht areas.

The reality is that keeping a fixed appointment at a hospital or visiting a GP, for example, requires advance planning and co-ordination for less-well-off rural dwellers to an extent that is outside the experience of most urban dwellers. In some instances this undoubtedly increases the stress that may be associated with the event and it may contribute to non-attendance with deleterious health affects. Simple activities such as shopping for convenience goods and attending church can involve complex advance planning and dependence on neighbours for the elderly in particular.

Other Public Services
The contraction or lack of provision of a range of other services in remote rural areas contributes to problems of accessibility with other negative implications. The increased incidence of burglary on the homes of the rural elderly accompanied by physical assault and, in some instances death, points to the vulnerability of people living alone in isolated locations, where garda patrols are infrequent, and undoubtedly contributes to stress. Recreational provision in rural Ireland has for long focused on a GAA pitch for football and/or hurling, with handball being a locally popular sport. The opportunities for involvement of girls in team sports, other than camogie in selected areas, is limited. The range of recreational provision has increased very markedly in towns in recent years to include playing courts and swimming pools, the costs of which are beyond the means of low density rural populations. Regular access to a wide range of recreational opportunities lies beyond the experience of many rural children because of the non-availability of local facilities and because of distance from towns.

The multiple weakening of service structures in rural areas characterised by low population densities and low income levels contributes to a weakening of social structures. The closure of shops and schools removes points where people meet, exchange news and discuss events of community interest. For the elderly, people on low incomes and women engaged in childminding in the home the loss of these points of contact contributes to isolation. The absence of a service structure reduces the employment base, removes a professional element who provided community leadership in the past and contributes to outmigration among young economically active people who desire parity of opportunity for their children with their peers in urban areas. Lack of accessibility to services is a multifaceted phenomenon which has far-reaching implications for other aspects of social and economic exclusion (Harvey, 1994).

OFFSETTING LOW LEVELS OF ACCESSIBILITY: PROVISION AND NEED

O'Shea (1996) argues rightly that equity considerations need to be incorporated within models of service provision and delivery which have hitherto focused almost exclusively on economic efficiency. Distance is one factor that creates inequities in accessibility, most notably for the less-well-off members of society. Distance merits inclusion in policy design for service delivery at a national level. Some recognition in this regard is present already, particularly in the case of the health services (Department of Health, 1993). There are also strategies in place involving state and voluntary agencies which are designed to overcome some of the barriers to accessibility associated with distance. Experience of using these services, as the evidence presented illustrates, reveals that difficulties of accessibility remain.

Transport initiatives have also been introduced to improve accessibility to services for rural dwellers. A postbus service has been operating very successfully in County Clare for several years providing transport in an area that lacks public transport provision. A dial-a-ride scheme centred on Kilkenny was introduced in 1996 (*The Irish Times*, April 29th, 1996). A subsidised transport service was initiated by Forum in north-west Connemara in 1994 (Taylor Lightfoot, 1995). Glennamaddy Social Services Council, with support from the Western Health Board, provides transport to a day-care centre in the village for the rural elderly and runs a meals-on-wheels service (O'Mahony, 1986).

Many other services are provided by voluntary agencies, local authorities and private business organisations which address problems associated with inaccessibility in rural areas (O'Mahony, 1985; Stuart and Cawley, 1991). Muintir na Tíre, with assistance from the Department of Health initially, and currently from the Department of Justice, runs a Community Alert scheme designed to increase security for the elderly in their homes (Muintir na Tíre, 1996). Meals-on-wheels are supplied in selected areas by voluntary groups with Health Board funding. A mobile library service is funded by some county councils which provides borrowing facilities for rural dwellers from village stopping points. Mobile banking services are available in some areas where fixed banking facilities are not present. Privately-owned mobile shops operate in some remote areas, although the cost of the goods provided tends to be higher than in towns because of the additional transport costs involved. Consideration is being given to the potential offered by new computer-based communications technologies to provide services through telecottages (Telecottages Ireland, 1994). Pilot projects involving an integrated approach to the provision of public services are in place in western counties under the aegis of the Minister for Western Development and Rural Renewal (Carey, 1995). There is therefore a growing body of experience relating to the use of non-conventional strategies to increase accessibility to services. This experience needs to be taken into consideration in designing remedial strategies in a comprehensive way.

In seeking to identify strategies that would help to ameliorate the problems of low accessibility experienced by the rural poor, it is recognised that no simple solutions exist. From reviewing the context within which accessibility has declined, some of the problems that have been identified through empirical research and some of the remedial approaches that have been adopted to date, it seems pertinent that the following should be taken into account:

1. The distance that must be travelled to deliver services to or to access services from rural areas creates burdens in terms of the expenditure of time and money. Provision needs to be made to help reduce these burdens and their negative effects for the less-well-off so that they are not excluded from services that are essential for personal health, social well-being and economic opportunity. Apart from the individual deprivation that results, additional costs may be incurred in the longer term for rural society and for health and education budgets, to pick two obvious examples. Formal recognition of the barrier that distance presents to the poorer sectors of rural society and devising methods to compensate for that barrier need to be incorporated into public policy design for service provision.

2. Experience of delivering services in rural areas to date illustrates that many agencies are involved including the EU, national government and state agencies, regional boards, local authorities, interest groups, private companies and voluntary organisations. These operate at a variety of geographical scales and co-ordination between agencies and across geo-

graphical areas is necessary. It is important that provision is made for inter-group and inter-area collaboration and co-ordination. Effective action at local level requires such co-ordination at higher policy and planning levels.

3. Flexible transport provision is of particular importance at a local level in offsetting inaccessibility because of the need to travel to centralised points to use many services. Attention needs to be given to devising transport strategies that provide accessibility to the widest range of services, for the largest number of people in need of those services, in the most economically efficient way possible. Options for mobilising services also merit attention.

4. Some aspects of service inaccessibility can be identified through objective research and inventories should be undertaken in this regard to inform policy. There are other aspects of inaccessibility which are related to the perceptions and preferences of users. It is essential that users' views are sought to ensure that remedial strategies meet the real needs that exist.

REFERENCES

Barrett, S. (1991) *Transport Policy in Ireland in the 1990s*, Gill and Macmillan, Dublin.

Batten, D. and Wiberg, U. (1988) *How to Reduce Disparities in the Provision of Services between Urban and Rural Areas: Examples from Sweden*, University of Umea, Sweden.

Breen, R. (1991) *Employment, Education and Training in the Youth Labour Market*, General Research Series Paper No. 152, Economic and Social Research Institute, Dublin.

Byrne, A. (1991) *North-west Connemara: A Baseline Study of Poverty*: Forum, Letterfrack.

Carey, D. (1995) Speech by the Minister of State, Donal Carey TD, at the fourth and final seminar on the theme "Rural renewal: an integrated approach to the provision of public services", Ennis, June 24, 1995.

Cawley, M.E. (1986) "Disadvantaged groups and areas: problems of rural service provision", in Breathnach, P. and Cawley, M.E. (eds.), *Change and*

Development in Rural Ireland, Geographical Society of Ireland, Special Publications, No. 1, St. Patrick's College, Maynooth.

Cawley, M.E. and Stevens, F.M. (1987) "Non-attendance at outpatient clinics at the Regional Hospital", Galway, Ireland, *Social Science and Medicine*, 25(11), 1189-96.

Cork County Council and FÁS (1994) *West Cork Island Study: Draft*, Cork County Council and FÁS, Cork.

Curry, J. (1980) *The Irish Social Services*, 1st. ed., The Institute of Public Administration, Dublin.

Curry, J. (1993) *The Irish Social Services*, 2nd. ed., The Institute of Public Administration, Dublin.

Department of Communications (1985) *Transport Policy*, Stationery Office, Dublin.

Department of Education (1995) *Charting our Education Future*, Stationery Office, Dublin.

Department of Health (1993) *Shaping a Healthier Future*, Stationery Office, Dublin.

Department of Health (1995) *Health Statistics 1993*, Stationery Office, Dublin.

Fyfe, G. ed. (1994) *Poor and Paying for It*, Scottish Consumer Council, HMSO, Edinburgh.

Harvey, B. (1994) *Combating Exclusion: Lessons from the Third EU Poverty Programme in Ireland 1989-1994*, Combat Poverty Agency, Dublin.

Interdepartmental Co-Ordinating Committee on Island Development (1996) *Report*, Stationery Office, Dublin.

Irish National Teachers' Organisation (1995) *Poverty and educational disadvantage; breaking the cycle*, INTO, Dublin.

The Irish Times, February 9, 1991, Widespread unease at An Post plan, p. 2.

The Irish Times, January 9, 1995, INTO wants aid for students in areas of poverty, p. 4.

The Irish Times, April 29, 1996, On-call buses to get rural areas moving, p. 4.

McCashin, A. (1982) "Social policy", 1957-82, *Administration*, 30, 203-24.

McKinsey International Incorporated (1979) *The Transport Challenge*, Stationery Office, Dublin.

McMahon, S. (1989) *Implications of School Consolidation for Rural Communities*, Thesis submitted in partial fulfilment for the Degree of Master of Rural Development, University College Galway.

Muintir na Tíre (1996) *Community Alert in Action*, Muintir na Tíre, Tipperary.

O'Connor, J., Ruddle, H., O'Gallagher, M. and Murphy, E. (1988) *Caring for the Elderly, Part II. The Caring Process: A Study of Carers in the Home*, National Council for the Aged, Report No. 19, Dublin.

O'Mahony, A. (1995) *Social Need and the Provision of Social Services in Rural Areas: A Case Study for the Community Care Services*, Socio-economic Research Series No. 5, An Foras Talúntais, Dublin.

O'Mahony, A. (1986) *The Elderly in the Community: Transport and Access to Services in Rural Areas*, National Council for the Aged, Report No. 15, Dublin.

O'Shea, E. (1996) "Rural poverty and rural service provision", 1996, in press.

Organisation for Economic Co-operation and Development (1991) *New ways of Managing Services in Rural Areas*, OECD, Paris.

Royle, S. and Scott, D. (1996) "Accessibility and the Irish islands", *Geography*, 81(2), 111-19.

Sixsmith, J. (1995) *Irish Women's Attitudes to Breast Cancer and Mammography: An Urban-Rural Comparison*, Thesis submitted for the Degree of M.A. in Health Promotion, University College Galway.

Storey, D. (1994) "The spatial distribution of education and health and welfare facilities in rural Ireland", *Administration*, 42(3), 246-68.

Stuart, M. and Cawley, M. (1991) "The voluntary sector and social service provision: the experience of Clare Social Services Council", in Varley, T., Boylan, T.A. and Cuddy, M.P. (eds.) *Rural Crisis: Perspectives on Irish Rural Development*, Centre for Development Studies, University College Galway.

Taylor Lightfoot Transport Consultants (1995) *North West Connemara Community Transport Study*, Forum, Letterfrack.

Telecottages Ireland (1994) *Newsletter*, 1, May 1994.

Western Development Partnership Board (1996) *Challenge: A Positive Future Through Action*, Western Development Partnership Board, Sligo.

Western Health Board (1995) *Community Care Annual Report for the Year Ended 31 December 1994*, Western Health Board, Galway.

8

RESIDENTS' PERSPECTIVES OF RURAL LIVING CONDITIONS IN CORK AND KERRY

David Storey
Geography Department, University College Worcester

INTRODUCTION

In the past studies of living conditions have tended to rely on the use of supposedly "objective" indicators chosen by researchers. Within geography, territorial social indicators were used to examine spatial variations in living conditions at a macro-scale (Smith, 1973; Knox, 1975). While this mapping of human wellbeing may well have its merits, it suffers from a number of limitations. It has been argued that such work amounts to "standardised abstraction" (Myers, 1987, 121) remote from the reality of peoples' lives. Much recent work, such as that by Cloke *et al.* (1995) in England and Wales, has tended to focus on the "subjective" views of local residents. This is seen by some as a more appropriate mode of investigation. It has been argued that to do otherwise "would mean failing to examine deprivation in the context that really matters: the life of the deprived person" (Eyles, 1987, 221).

This chapter explores living conditions in rural Cork and Kerry by comparing observed material circumstances with the expressed views of residents. This analysis enhances the construction of a more complete picture of rural living conditions, based not solely on the presence or absence of selected items, but taking into account the subjective views of the people themselves. In this way there is a bringing together of material and experiential issues as advocated by Cloke *et al.* (1995).

Central to the study is a consideration of the extent to which people in rural areas may be said to experience deprivation. Much debate has occurred over the precise meaning of terms such as poverty and deprivation with a general agreement that relative conceptualisations appear to be the most appropriate, at least within so-called "developed" societies. (Holtermann, 1975; Holman, 1978; Townsend, 1979, 1987; McLaughlin, 1981; Piachaud, 1987; Ringen, 1988; Callan *et al.*, 1989). Such an approach considers living conditions in relation to societal norms rather than to some absolute standard. This allows judgements to be made on the basis of what is "normal" in the society under consideration, rather than on the basis of a fixed idea of what is "sufficient" or "necessary". Ideas of what comprises a decent lifestyle change over time, therefore what constitutes poverty will also change over time.

In order to assess material well-being, a "style of living" approach is used (Townsend, 1979; Mack and Lansley, 1985; Economic and Social Research Institute/Combat Poverty Agency, 1988). This involves assessing the conditions of individual households by examining whether or not they possess certain material items. Following this assessment, the paper focuses on respondents' perceptions. An exploration of whether or not respondents believe themselves to be poor is followed by an examination of their views on, firstly, what constitutes poverty, secondly, whether poverty and deprivation exist, either in Ireland generally or in their own area, and, finally, what they perceive to be the causes of poverty. In all these instances the views of respondents are examined in relation to their observed material circumstances, as evidenced through the "style of living" indicators, in order to establish whether attitudes vary due to differing material conditions.

The analysis is based on the results of a questionnaire survey carried out in the District Electoral Divisions (DEDs) of Castletown, Barnacurra and Tahilla in counties Cork and Kerry (Figure 8.1). These three areas contain no major settlements and, in keeping with much of west Cork and Kerry, all three lost population throughout the 1980s. All occupied households in the study areas were surveyed on a "door to door" basis. A total of 191 successful responses were obtained. Interviews were conducted with one or more adult residents. (While the household is the unit of

Residents' Perspectives of Rural Living Conditions

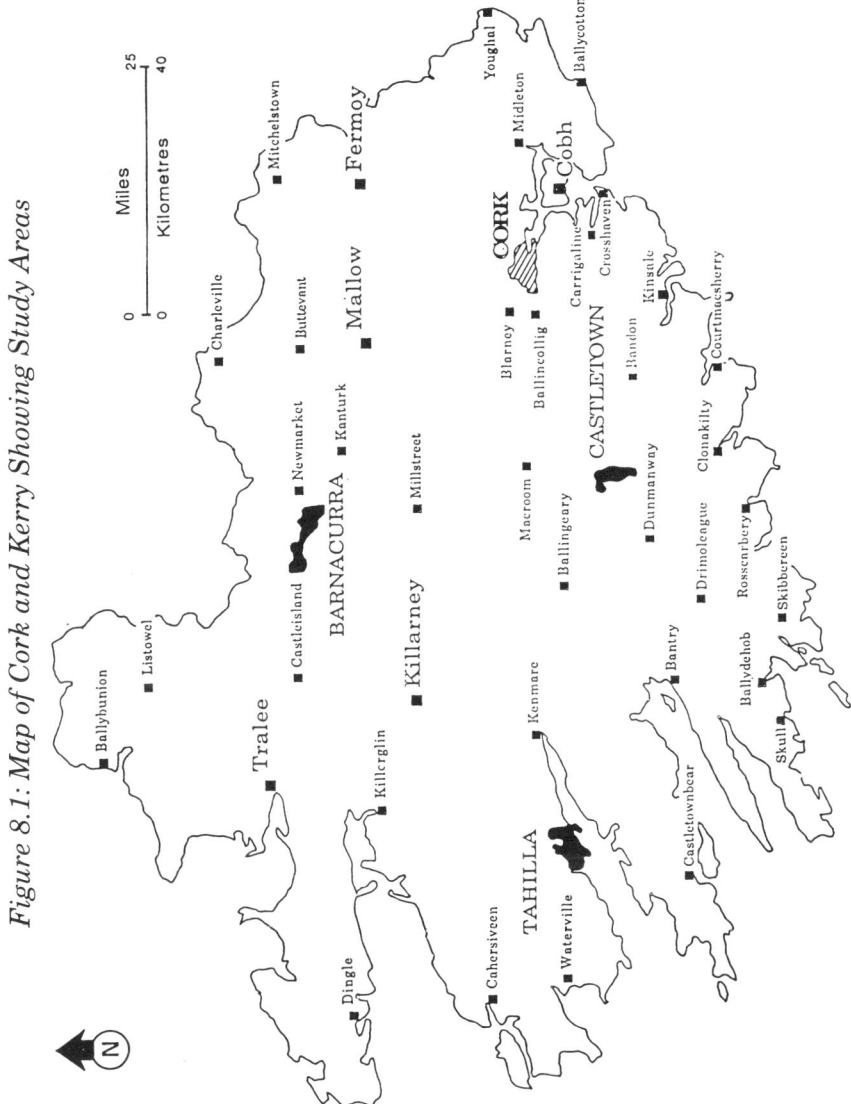

Figure 8.1: Map of Cork and Kerry Showing Study Areas

analysis, it is acknowledged that differentials in life quality may well occur between members of the same household.) Although differences exist between the localities, the intention here is not to highlight the uniqueness of the areas under study. Rather, it is an attempt to utilise differing methodological perspectives in the exploration of rural living conditions.

HOUSEHOLD CONDITIONS

Using a list of ten household items based on those employed by Callan *et al.* (1989) and Byrne *et al.* (1991) respondents were asked if they possessed each of the items. The results clearly show that there is a significant minority of households who lack many of the accepted features of modern life in Ireland, including basic sanitary facilities (Table 8.1). While a large number of households possess all ten items, there are many who experience a "multiple lack" and some who are lacking the majority of items (Figure 8.2).

Table 8.1: Non-possession of Selected Household Amenities in Study Areas

Item	Households Lacking Item (%)	Households Having Enforced Lack (%)	"Lacking" Households whose Lack is Enforced (%)	Total Number of Households Lacking Item
Electricity	2.1	1.6	75.0	4
Running water	8.9	8.0	88.2	17
Toilet	13.1	12.0	92.0	25
Bath/shower	15.7	13.6	86.7	30
Television	9.0	4.8	52.9	21
Video recorder	52.7	40.8	51.5	126
Refrigerator	3.7	2.6	71.4	7
Wash. machine	25.5	14.9	62.8	48
Telephone	36.2	14.9	41.2	68
Central heating	40.4	21.3	52.6	76

Figure 8.2: Number of Amenities Lacked

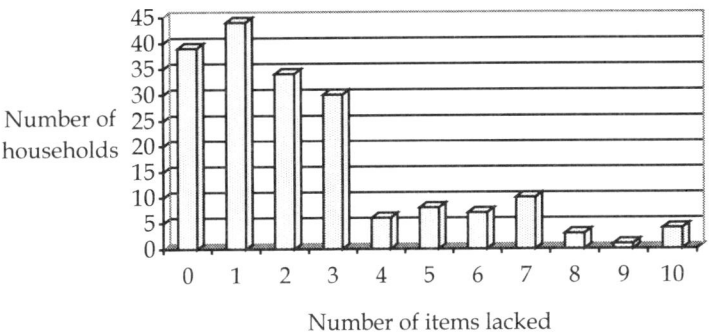

Non-possession of a particular item is only a problem, it might be argued, if a household actually wants that item. This means it is necessary to distinguish between "voluntary lack" and "enforced deprivation". In order to do so respondents were asked to indicate, in respect of each item which they did not possess, the reason for non-possession. The results indicate that in the case of the basic sanitary facilities lack through choice is not a significant explanatory factor. Absence appears almost exclusively due to financial constraints (Table 8.1). In the case of the other household items choice does appear to enter into the reckoning when dealing with the absence of these items from some households. However, it must be borne in mind that "an individual's perception of whether or not they are exercising choice will itself depend on the extent to which they are deprived" (Mack and Lansley, 1985, 34). There is a large number of households where no reason was given for the absence of some of the items. In 28 of 38 such cases, the household lacks more than three items and 22 of them lack more than five. Thus, it is mainly the materially poorer households which provide no reason for non-possession. In such households it might be reasonable to assume that, given the probable existence of other more pressing needs, the idea of possessing the item has never even been considered. A household may have opted, for instance, to purchase a refrigerator instead of a washing machine. Under these circumstances, the absence of the washing machine cannot be seen as the result of an unconstrained choice. Rather, it should be seen as an enforced lack. This effectively means that a large proportion of those not stating a reason for their non-possession may well be unable to afford the item even if they did want it. Their economic circumstances imply that their ability to choose is highly limited.

A significant level of material deprivation exists in these localities. Although some households have decided they do not want or do not need particular appliances, such apparently unconstrained choice does not explain the lack of basic sanitary items in households and falls well short of fully explaining the lack of other appliances. In addition, due to constraints on household choice, it seems quite probable that the extent of enforced lack is much greater than that actually recorded. The analysis thus provides a measure of the minimal amount of involuntary lack. The true level of enforced deprivation is almost certainly somewhat greater.

RESPONDENTS' ASSESSMENTS OF THEIR LIVING CONDITIONS

Respondents were asked if they perceived themselves to be poor. Thirty one per cent of households regard themselves as being poor, while less than one third of households claim never to have been poor. The results strongly suggest that whatever stigma may attach to being poor, many people are nevertheless willing to admit to living in such a state. Whatever "psychic income" (Cloke and Davies, 1992, 355) may be said to derive from rural living clearly does not offset poor material conditions for many people. This may result, in part at least, from the well documented history of rural disadvantage in Ireland.

A comparison of the "objective" situation with peoples own subjective views shows that there is a clear relationship between a lack of household items and a perception of being poor (Figure 8.3). Of those who might be "objectively" deemed less well-off, the majority have no difficulty in labelling themselves as poor. Many people characterised their position as one of "existing", "struggling" or "just carrying on". Despite this, some materially-deprived households do not regard themselves as poor. This may be explained by the fact that respondents' views are obviously conditioned by their own experiences. Many elderly more impoverished residents may have memories of even greater deprivations during their childhood and earlier stages of their lives. This will temper their view of their current state and, as a consequence, they may be quite content with their lot, viewing it as far superior to what they once endured. Their judgement is based on a comparison within their own lived experience and not on contemporary standards. They may have less materialistic goals and, as a result, may be satisfied with less. Thus, although it may well be argued that they have been denied many of the commonly perceived benefits of modern society, they may be reasonably content with their present conditions. There is also a suggestion of denial for psychological reasons. Thus, in the words of one respondent: "if you admit to poverty you get depressed — you have to have hope".

Figure 8.3: Self-defined Poor Classified by Number of Amenities Lacked

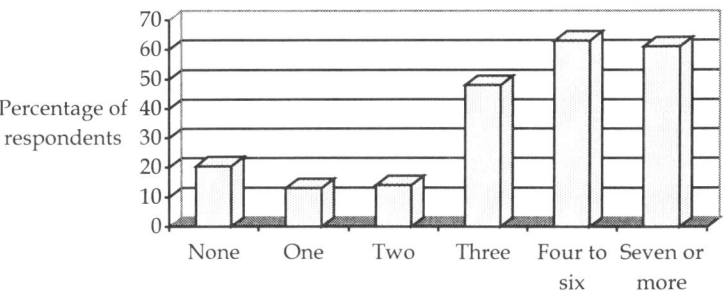

Five of the households who possess all the selected items categorise themselves as poor. This again could be taken as evidence of the setting of a relative standard of poverty by the respondents. They may be less well-off than others they know and they may feel that their possessions represent the minimum which any household should have in Irish society. Equally, their current circumstances may represent a decline from an earlier period in their lives when they were able to purchase the items concerned. A reduction in income, or some other unforeseen change in circumstances, may have precipitated a decline in well-being.

There appears to be a disproportionate degree of vulnerability amongst older non-family households. One third of those who regard themselves as poor are one-person households and 53 per cent are either one or two person households. Most of these are middle-aged or elderly people. While this suggests that rural deprivation is a particular problem for older households, this should not obscure the fact that, with the exception of those in the pre-family and early school categories, there are households at virtually all stages in the family cycle who claim to be poor (Table 8.2). It is also apparent that the self-defined poor are spread throughout the socio-economic spectrum and are not confined to any one group (Table 8.3). While farmers appear more vulnerable (35 per cent of the total claiming to be poor are farmers) it is clearly incorrect to conclude that poorer living conditions are the preserve of a particular occupational group. The unremunerative nature of small farming would appear to be borne out by the evidence presented here. In addition to those currently employed, 28 per cent of those who are retired, a category composed of many ex-farmers, and 50 per cent of the unemployed claim to be poor.

Table 8.2: Percentage of Family Units in Each Category Claiming to be Poor

Stage in Family Cycle	Percentage of Families
"Pre-Family" stage (both partners under 45 years of age with no children)	0.0
"Pre-School" stage (oldest child is aged 0-4 years)	26.7
"Early School" stage (oldest child is aged 5-9 years)	0.0
"Pre-Adolescent" stage (oldest child is aged 10-14 years)	16.7
"Adolescent" stage (oldest child is aged 15-19 years)	36.8
"Adult" stage (oldest child is aged 20 years or over)	35.4
"Empty Nest" stage (both partners aged between 45 and 64 years with no children resident)	25.0
"Retired" stage (both partners aged 65 years or over with no resident children)	23.1

Table 8.3: Percentage in Each Socio-economic Group Claiming to be Poor

Socio-economic Group	Per centage of Households
Agricultural occupations	27.6
Professionals	0.0
Employers and managers	11.1
Salaried employees	0.0
Intermediate non-manual	44.4
Other non-manual	12.5
Skilled manual	13.3
Semi-skilled manual	0.0
Unskilled manual	80.0

It is clear that different groups in the study areas experience rural living somewhat differently. While many people experience deprivation, many others are well-off. This differential experience in itself may well serve to increase the problems experienced by the less well-off giving rise to feelings of further isolation. It can be concluded that there is a considerable degree of correspondence between so-called "objective" assessments of living conditions and the subjective views of respondents, with the majority of those deemed deprived according to the "objective" analysis regarding themselves as poor. However, this conclusion must be qualified by the fact that some households lacking a number of the selected amenities do not regard themselves as poor, while others who possess most of the items under consideration do regard themselves as poor. Clearly, people have differing conceptions of need and life in rural Ireland impacts differently on different households.

RESPONDENTS' ATTITUDES TO THE EXISTENCE OF POVERTY
It is clear that most people are aware of poverty at a national level. However, the majority of respondents do not see it as a local or as a rural issue. There is a failure to relate it to their own environment. While over 80 per cent of respondents think poverty is a problem in Ireland, less than half see it as a problem in their own

locality. Many respondents referred to poverty as a phenomenon occurring in the cities but not in rural areas. Poverty is seen to exist elsewhere but not in their midst. This adds to the notion of rural poverty as a phenomenon more "hidden" than its urban counterpart.

Some respondents commented to the effect that poverty was no longer a problem in Ireland but was restricted to the "Third World", a response also noted by Byrne *et al.* (1991) in Connemara. This suggests a view of poverty as a condition where people don't have enough to eat. This absolute definition implies that poverty does not exist in Ireland since people will always have enough to eat (or so we believe). A contrasting view was expressed by one respondent in the present survey: "we have Ethiopia at home".

Those who are materially better-off are less likely to view poverty as a problem in their own locality (Figure 8.4). Similarly, just over 63 per cent of those who regard themselves as poor think there is a poverty problem locally, with only 23.6 per cent of those who don't thinking there is. There is a clear suggestion here that those who themselves have no direct experience of poverty are less likely to be aware of its existence in their midst. In the words of one respondent, "the rich man thinks you have enough".

In summary, it appears there is widespread recognition of the existence of poverty in Ireland generally, but much less acceptance of its existence within respondents' own localities. There is a strong indication that those who have no direct experience of poverty are much less likely to view it as a problem locally. For those who are somewhat better off, the conditions experienced by some of their neighbours are hidden. This is particularly so in relatively sparsely populated areas such as those under consideration here. The lack of any overt external manifestation of poverty and deprivation in rural areas may provide a partial explanation for this. Unless one is inside the home, extreme deprivation may not be apparent. In a situation where there may be little or no contact between those who are deprived and the rest of the local population, the conditions under which the former live may remain well hidden. This lends further support to the idea that one of the chief problems surrounding rural deprivation is the extent to which its existence is denied.

Figure 8.4: Poverty Perceived as Problem Classified by Amenities Lacked

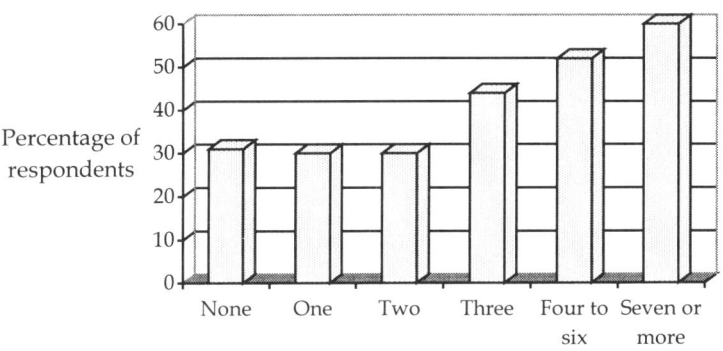

RESPONDENTS' VIEWS OF THE CAUSES OF POVERTY

There is a plurality of views among respondents concerning the causes of poverty. Many responses reflect a "victim-blaming" syndrome; a stance which attributes causality to individuals themselves or to their characteristics or behaviour. The most oft-quoted response to the question of what causes poverty is laziness (Table 8.4). The total figure of 21.9 per cent compares with the 21 per cent reported by Byrne *et al.* (1991) for Connemara and the 25 per cent of Riffault and Rabier (1977) nationally in the 1970s. Thus, it appears that little has changed in this respect with the image of the poor as being too lazy to work still very real for one in five people.

While undoubtedly less popular, other responses based on the perceived behaviour of the poor themselves were also cited. These include poor education and bad financial management. This argument is predicated on the belief that the poor would not be poor if they managed their money better. It is supposed by some that they need to be better educated in order to be able to do so, through learning to shop around for cheaper food and adopting other cost-saving devices. Such reasoning appears flawed given that the ability to survive on a low income appears to reflect good, rather than bad, financial management. More importantly, improved financial management, even if such is possible, does not alter the actual household income and does not serve to change the conditions necessitating the improved management in the first place. Suggestions that the poor need to alter their spending patterns and general money management techniques represent a response to poverty rather than an explanation of it.

Table 8.4: Causes of Poverty Cited by Respondents

Cause	Total
People are lazy	21.9
People have been unlucky	20.2
Lack of jobs	19.7
Poverty is inevitable	18.0
There is injustice in society	15.3
Lack of education/poor financial management	9.6
Government's fault	8.5
Drinking/smoking	8.0
Low pay/high cost of living	6.9
"Keeping up with the Jones's"	3.7
Home environment/circle of poverty	3.2
Too many handouts	2.7
People unwilling to help themselves	1.6

Drinking and smoking are cited by some households in all three areas. This view is considerably less popular than in previous research (Riffault and Rabier, 1977; McAirt, 1979). The results suggest that this idea of profligate expenditure on the part of the poor may be receding. An indication of this is revealed in the comment of one respondent: "drink — but maybe they are entitled to it". Mack and Lansley's (1985) criticism of smoking as a cause of poverty is pertinent. Their survey in Britain indicated that the majority of those who smoked would still be in poverty even if they gave up the habit. The financial saving would not be sufficient to raise their income above the poverty line.

A small number of households cite the phenomenon of "keeping up with the Jones's" as contributing to poverty. This can be imputed as overspending by households ("too much high living") and thus falls into the "victim-blaming" category. While some households may well live beyond their means in order to keep apace with their neighbours, it could be argued that this is in itself indicative of an unequal society and reflects the extent to which people are encouraged to obtain more goods and services. Some households cite home or family reasons as causes of poverty and in some cases people were seen as "unwilling to make an effort to improve". If all these individualist causes are taken together, it is apparent that a sizeable proportion, 44.2 per cent, of households in all three areas feel people themselves are to blame for their poverty. Thus, there is a considerable amount of "victim blaming" in the attribution of causality.

Just over one-fifth of all respondents attribute the existence of poverty to the fact that people have been unlucky. In addition, a sizeable proportion of households see poverty as inevitable in modern society. These could be categorised as fatalistic views of "the poor will always be with us" variety. When those who regard poverty as due to bad luck are added to those who see poverty as inevitable in a composite "fatalistic" category, 31.6 per cent could be said to adopt such a view.

The third most frequent response to the cause of poverty is the lack of a job. This is cited by approximately one fifth of interviewees. With relatively high levels of unemployment nationally, it is somewhat surprising that the figures are so low. This could be attributable to the fact that unemployment rates in the study areas, as in rural Ireland generally, are relatively low. The fact that poverty may be seen to exist among those, such as small farmers, who are working may serve to remind residents that it is not restricted to those who are unemployed. Related to this, low pay and the cost of living is seen by some as a causative factor.

The recognition, by some, of issues of social justice is evident. The state is not felt to be doing enough to resolve the problem. The figures are considerably lower than the 47 per cent in the Connemara survey who cite "bad government planning" as a cause of poverty (Byrne *et al.*, 1991). However, if these proportions are added to those who feel lack of jobs, the high cost of living and the existence of injustice are to blame for poverty, then the proportions of respondents subscribing to a view which sees poverty as determined by factors beyond the control of the individual and concerned with structural conditions within society are almost identical in both Connemara (47 per cent) and Cork/Kerry (44.6 per cent).

The complexity in arriving at an understanding of poverty and deprivation is emphasised by the fact that many respondents express views which fall into various combinations of individualist and structuralist as well as fatalistic causes. (Thus the figures in Table 8.4 total over 100 per cent). If those who used a combination of two or more of these categories are excluded then 115 responses fit solely into one category. Over 44 per cent of these adopt structural explanations, 35.7 per cent adopt individualistic explanations and 20 per cent adopt fatalistic explanations. Thus, structuralist explanations appear to hold more sway than others.

It is apparent that those who are materially better off are more prone to adopt a "victim-blaming" stance and are much less likely to attribute causality to structural factors than are those materially less well-off (Table 8.5). Structural explanations are favoured by those who are themselves poor by their own admission. While there is a degree of ambiguity in the results, there is, nevertheless, a strong suggestion from the evidence here that "blaming the victim" is more prevalent among the non-victims, with the less well-off more likely to adopt structuralist explanations. Those who are poor are more likely to be aware of the negative impacts of external processes than are those who have never experienced poor conditions.

In summary, attitudes to the causes of poverty are quite varied. However, if only those responses which fit unambiguously into one category are examined, structuralist views are the most popular. People may now be more willing to accept that poverty is not due to laziness and the idea of the "idle dole sponger" appears somewhat less popular than it was at one time. Nevertheless, there is still a considerable degree of support for the idea that the poor are themselves to blame for the predicament in which they find themselves. This view appears to be quite prevalent among better-off households.

Table 8.5: Respondents' Perceptions of Causes of Poverty

Household Circumstances	Individualist	Structuralist
Possess all ten items	61.3	38.7
One item lacked	46.7	53.3
Two items lacked	30.4	69.6
Three items lacked	45.0	55.0
More than three items lacked	27.8	72.2
Regard themselves as poor	20.7	79.3
Regard themselves as non-poor	55.6	44.4

CONCLUSIONS

In the context or England and Wales, Cloke *et al.* (1995) ask, amongst other things: Do different people experience the "same" rural world differently? Do normative indicators coincide with expressed views of the problems experienced by rural dwellers? Are peoples experiences of these problems affected by differing expectations, fear of stigma or related factors? The evidence presented here suggests that the answer to all three questions is yes. Firstly, people *do* experience the rural world somewhat differently. There is a diversity of living conditions in rural Ireland. The results highlight the extent of relative deprivation and underline the disparities in living conditions which exist within particular localities. They point to the incidence of deprivation amongst particular groups such as small farmers and the elderly (although such conditions are not unique to those groups). Materially well-off households live in close proximity to much poorer households. Many lack a range of commonly accepted household items while conditions are considerably better for many others. Secondly, these differences in material conditions are reflected in the expressed opinions of respondents. There is a reasonably close correspondence between households deeming themselves poor and those lacking a number of household facilities. In other words there is a strong link between what might be termed "objective" criteria concerning rural living conditions and the more "subjective" views of residents in rural areas. There appears to be a relationship between the material and the experiential.

Finally, there are attitudinal differences between those who can be deemed to be materially better-off and those who are less well-off. The latter are more likely to view poverty as a problem in their own locality and to attribute its existence to structural factors. Those who are better-off, on the other hand, tend not to regard poverty as a problem in their own area and to attribute it to individualist or "victim-blaming" causes. Thus, there appear to be links between peoples material circumstances and their expressed views on poverty and deprivation. This is not to ignore differences in responses between people in similar material circumstances. This may be due a variety of factors. For some a stigma may well be attached to the condition of poverty. The setting of relative standards may also be important. People may view their current circumstances as an improvement on more harsh conditions experienced in the past. Thus, in some cases judgements are made on the basis of past standards not existing societal norms.

The study highlights the diversity of living conditions in rural Ireland and suggests that a significant number of rural residents experience conditions of deprivation. There appears to be a strong link between observed material conditions and expressed views. Household living conditions would certainly appear to play a role in shaping general views on poverty and deprivation with the better-off more likely to endorse "victim-blaming" perspectives, while the less well-off appear more aware of the influence of factors external to the individual or household.

REFERENCES

Byrne, A., Laver, M., Forde, C., Cassidy, L., Keane, M. and O Cinneide, M. (1991) "North-West Connemara: A Baseline Survey of Poverty", FORUM, Letterfrack.

Callan, T., Hannan, D.F., Nolan, B. and Whelan, B.J. (1989) "Measuring poverty in Ireland: a reply", *Economic and Social Review* 20(4), 361-368.

Callan, T., Nolan, B., Whelan, B.J., Hannan, D.F. and Creighton, S. (1989) *Poverty, Income and Welfare in Ireland*, The Economic and Social Research Institute, Dublin.

Cloke, P.J and Davies, L. (1992) "Deprivation and lifestyles in rural Wales – I. Towards a cultural dimension", *Journal of Rural Studies* 8(4), 349-358.

Cloke, P., Goodwin, M., Milbourne, P. and Thomas, C. (1995) "Deprivation, poverty and marginalization in rural lifestyles in England and Wales", *Journal of Rural Studies* 11(4), 351-365.

The Economic and Social Research Institute/Combat Poverty Agency (1988) *Poverty and the Social Welfare System in Ireland*, Combat Poverty Agency, Dublin.

Eyles, J. (1987) "Poverty, deprivation and social planning", in *Social Geography: Progress and Prospect*, Pacione. M. (ed.) pp. 201-251, Croom Helm, London.

Holman. R. (1978) *Poverty. Explanations of Social Deprivation*, Martin Robertson, Oxford.

Holtermann, S. (1975) "Areas of urban deprivation in Great Britain: an analysis of 1971 census data", *Social Trends* 6, 33-47.

Knox, P.L. (1975) *Social Well-Being: A Spatial Perspective*. Oxford University Press, London.

McAirt, J. (1979) "The causes and alleviation of Irish poverty", *Social Studies* 6(3), 255-264.

McLaughlin, B.P (1981) "Rural deprivation", *The Planner* 67(2), 31-33.

Mack, J. and Lansley, S. (1985) *Poor Britain*, George Allen and Unwin, London.

Myers, D. (1987) "Community-relevant measurement of quality of life: a focus on local trends", *Urban Affairs Quarterly* 23(1), 108-125.

Piachaud, D. (1987) "Problems in the definition and measurement of poverty", *Journal of Social Policy* 16(2), 147-164.

Riffault, H. and Rabier, J.R. (1977) "The Perception of Poverty in Europe", Commission of the European Communities, Brussels.

Ringen, S. (1988) "Direct and indirect measures of poverty", *Journal of Social Policy* 17(3), 351-365.

Smith, D.M. (1973) *The Geography of Social Well-being in the United States: An Introduction to Territorial Social Indicators*, McGraw-Hill, New York.

Townsend, P. (1979) *Poverty in the United Kingdom: A Study of Household Resources and Standards of Living*, Penguin, Harmondsworth.

Townsend, P. (1987) "Deprivation", *Journal of Social Policy* 16(2), 125-146.

Part 3

URBAN POVERTY

9

DECONSTRUCTING URBAN POVERTY

Andrew MacLaran
Department of Geography, Trinity College Dublin

INTRODUCTION

As in warfare, final goals should determine strategy and strategy determines tactics. So, before we decide how we do things, should we not be clear about the goals of the poverty industry? For example, does it aim to eliminate poverty, or merely to alleviate its most visible evidence? Is it about managing the poor, co-opting them and diverting overt conflict? Is it concerned more for the eradication of those elements associated with poverty which impinge adversely on the rest of society, such as particular types of crime, or are we simply trying to present the appearance that something is actually being done in order to create an impression that society really does care? Or is its aim merely the creation of employment for those engaged in the poverty industry itself?

PERSPECTIVES ON THE NATURE OF POVERTY

The answer to this seemingly simplistic question depends primarily upon one's conception of poverty and its causes. Most have probably by now abandoned the concept of absolute poverty, the idea of poverty as a fixed and measurable datum. It is an approach which can be traced to the nineteenth century. One establishes the costs of a fixed minimum level of food, clothing, shelter and other necessary items (down to haircuts and shoe laces) and determines whether a household's income suffices to meet this poverty line. Then, if one does have sufficient for the approved bare necessities but fecklessly spends some resources on a luxury or two, one falls into "secondary poverty". Poverty is now rarely defined in this way, at least in the developed world. Even for a single item in the diet there is no determinable datum. For example, during the 1960s, while the US National Research Council recommended a daily in-take of 70 mg/day of vitamin-C, the British Medical Association was content to recommend 20 mg. But if one cannot establish a datum for a single nutritional requirement, what possibility is there of doing so for other elements of human need? Adam Smith promoted a relativistic conception of poverty in *The Wealth of Nations*:

> By necessities I understand, not only the commodities which are indispensably necessary for the support of life, but by whatever the custom of the country renders it indecent for creditable people, even of the lowest order, to be without.

and this sentiment is expressed by an even earlier sage:

> Paupertas enim est non quae pauca possidet, sed quae multa non possidet (Antipater).
>
> (Poverty does not only mean the possession of little, but also the non-possession of much.)

While many might not ascribe to the analyses of Karl Marx, one might at least find a measure of agreement with him when he opines that:

> Our needs and enjoyments spring from society; we measure them, therefore, by society and not by the objects of their satisfaction. Because they are of a social nature, they are of a relative nature.

It is possible to cite a range of dimensions of life which are important contributors to levels of living: health, financial status, education, housing, character of employment, family stability, neighbourhood environment and security, quality of our leisure, access to amenities, ability to participate in and have control over the society in which we live. All represent facets of well-being which research has shown to be more or less important (see Knox and MacLaran, 1978). Many, though not all, are clearly dependent on one's financial situation. Although financial resources may permit one to purchase health care services, even the wealthy can be deprived in terms of important facets of life such as health status or caring inter-personal relationships. Most of us do better, or worse, in terms of one or other of these domains of life. We may sacrifice leisure time to earn more money in order to have a better house in a better neighbourhood. There are some who perform well over most of the dimensions simultaneously; the multiply privileged. But equally, there are those who do badly with respect to a wide range of conditions — those who are multiply deprived. It is important to emphasise this by making an inclusive statement. "The poor" do not constitute an identifiably separate group which is identifiably different from the rest of the community in terms of their attitudes, assumptions, norms and expectations. They are simply those who are excluded from full participation in society, primarily because they are deprived in terms of the economic dimension of well-being.

So, again, what constitute the goals of an anti-poverty programme? If they were simply concerned with lifting income levels to above a certain datum (the fixed poverty line) this could be achieved by minimum wage legislation, a negative income tax and by transfer payments. There would be little need for a well developed poverty industry. On the other hand, if one agrees that poverty is a relative concept encompassing inequality, what then might constitute the goals of an anti-poverty programme? How much inequality is acceptable? If one cannot make explicit the strategic aims of such a programme, how can appropriate policies be devised?

At a second level of analysis, the strategies which we do adopt to combat poverty clearly depend upon what are to be regarded as the sources of such multiple deprivation. A range of explanations cite the manner in which poverty conditions may be passed from one generation to the next. These are well known to most: the stage in the life cycle, the inter-generational cycle of poverty, personal disability, the culture of poverty and the foundations of social structure.

1. The Life Cycle Concept
The financial situation of a household rises above and dips below what are deemed to be acceptable levels (the poverty line) according to the stage in the life cycle of the family. The relative prosperity of two earners in the pre-children phase gives way to relative economic deprivation as one earner withdraws from the labour force to bear and rear children, who add additional expense to the household budget. Greater prosperity as the first children bring in a wage or leave home subsequently gives way to a return to deprivation during old age and loss of income upon retirement.

2. The Cycle of Poverty
The children of parents on low wages may suffer from a poor diet, poor housing conditions and from frequent periods of absence from school due to illness. The home environment is also likely to be unconducive to learning, lacking books, space and quiet for studying. "Learned helplessness", reduced powers of concentration and low levels of determination to cope successfully with difficult tasks each contributes toward such children becoming low achievers, commanding few marketable skills, thereby gaining employment which is poorly remunerated and ending up as low-income parents in their own right.

3. Personal Disability
This stresses the inability of the individual to gain access to better paid employment as a result of personal factors such as mental or physical disability.

4. The Culture of Poverty
The culture of poverty promotes the concept that neighbourhood and parental norms in slum areas are deviant, that "success" is achieved outside the normal measures of success and failure and that these deviant norms are passed on from generation to generation, making it culturally difficult for the young to break out of their poverty and assimilate into wider society.

A certain element of veracity might be found in each of these explanations of the manner in which poverty conditions might be passed from one generation to the next. There may well exist unsurmountable cultural hurdles which militate against the children of the poor settling with ease into a middle-class office environment. Just as miners are likely to be drawn from mining villages within which each new generation of miners is socialised, so the next generation of Dublin's middle-class is now being socialised in areas like Foxrock and Killiney, the factory workers of the future being drawn largely from the city's working-class areas. There is, of course, some inter-generational mobility, but this is very limited (Hutchinson, 1969). However, such explanations conveniently throw the burden of causation onto the individual, failing to explain why such disparities in conditions of well-being exist in the first place. This brings us to a fifth possibility:

5. Structuralist Explanations

In contrast, structuralist approaches emphasise the role of wider society and its dominant structures. They emphasise the concept that poverty simply represents one end of a continuum encompassing the total distribution of income and wealth.

The distribution of economic well-being is seen to be determined primarily by whether one is an owner of capital[1] or merely a seller of labour power (if one is able to find a buyer at all!). The division of the social product between the returns to capital and to labour are determined as an outcome of class conflict; a bargaining process between employers and labour. In this process of conflict over the distribution of social product, the enhanced mobility of capital can be used to move away from militant workforces, labour power being far more tied to place due to its residential immobility. So, capital shifts according to the relative profitability of locations, internationally, regionally or even locally, perhaps involving factory suburbanisation in order to engage a more pliant and less organised labour force of female or part-time workers.

Under this concept, the primary determinant of income is the distribution of wealth or the ownership of capital. All investments are a charge on the future expenditure of labour power, such investments merely representing claims upon the product of future labour power:

- **Equity shares** in companies are claims to share in the profits created by future labour power working in private-sector companies.

- **Commercial rents** represent claims to a charge on the profits created by future labour power making use of leased land/buildings.

- **Gilt-edged stock** (government securities) are a claim on the future capacity of the government to raise taxation and pay for its borrowings.

Then, at a secondary level of structuration, the level of income which groups within the labour force can command depends upon their degree of control over monopolisable skills and their ability to organise and bargain collectively.

This structuralist view of poverty clearly comprehends relative deprivation as an inherent consequence of relative advantage, each representing different extremities of the same distribution.

MEASURING INEQUALITY

Statistics measuring the distribution of income and wealth are notoriously unreliable as a result of the functioning of a sizeable hidden economy, the evasion of taxes and death duties and the holding of wealth outside the jurisdiction. Additionally, wealthier groups are also far less likely to participate in voluntary surveys of household income and wealth ownership (Nolan, 1991). Nevertheless, some attempts have been made to measure inequality and such studies clearly show that we continue to live in an intensely unequal society (Nolan, 1991; Nolan, 1992; Nolan et al., 1994; Callan et al., 1996). International comparisons have revealed Ireland to be one of the most unequal of O.E.C.D. countries (Atkinson, Rainwater and Smeeding, 1995, 46), although Eurostat data suggest that its income distribution differs little from the E.U. norm (Eurostat, 1997). Presenting evidence for the U.K., Scott (1994, 107) finds that "while the top 20 per cent enhanced their position over the 1980s, all the remaining 80 per cent saw their position worsen" during a decade when Conservative governments presided over the redistribution of social product in favour of the better off.

In the early 1970s, the Royal Commission for the Distribution of Income and Wealth estimated that the richest 5 per cent of the population in the U.K. owned 54 per cent of all wealth and that 10 per cent of the population accounted for nearly 72 per cent. Meanwhile, the bottom 80 per cent shared just 24 per cent of all wealth. For Ireland, shortcomings in the available statistical sources for evaluating the distribution of wealth are even more serious than for Britain and have led to considerable debate about the degree of inequality present in the ownership of wealth in the country (Byrne, 1989). Lyons (1975) calculated that inequality in the distribution of wealth was considerably greater in Ireland than in Britain, which was confirmed by Harrison and Nolan (1975) who estimated that the richest 5 per cent account for around 70 per cent of all wealth with the bottom 65 per cent holding only 2 per cent, with Chesher and McMahon (1976) subsequently arguing that the top 5 per cent actually controlled 57 per cent of wealth with the bottom two-thirds of the population accounting for 10 per cent. More recent estimates based on household survey data show that while the richest 10 per cent own 42 per cent of all wealth, the bottom 50 per cent control just 12 per cent (Nolan, 1991).

However, although each of these such studies reveals the continuation of massive inequality in the distribution of wealth, the statistics upon which they are commonly based considerably understate the degree of inequality in the distribution of productive wealth. That is because the greater part of the "wealth" which lies within the control of the less well off exists in the form of unproductive use-values embodied in owner-occupied housing (see Nolan, 1991; Honohan and Nolan, 1993). Although such assets represent a store of realisable exchange values and have become a significant medium for the inter-generational transmission of wealth, such wealth differs very significantly from that type of productive wealth (i.e. shares, gilt-edged stock, investment property, etc.) which generates monetary income and which confers immense economic power upon the very rich. Thus, during the early 1960s, Meade (1964) estimated that because investment returns were higher for larger investors, just 5 per cent of the population pocketed 92 per cent of the income from investment.

One may balk at some of these ideas. After all, are we not now meant to be living in shareholding democracies? Widespread privatisation of public-sector businesses and utilities in the UK certainly attracted wider share-owning interest than had traditionally been the case. Moreover, even if we do not own shares individually, most probably do so indirectly through pension funds and the investments of life insurance companies. That may be true to a degree. However, Ambrose and Colenutt (1975) revealed that in the early 1970s some 85 per cent of the returns from life assurance and pension fund investment went to people within the top 20 per cent band of income in Britain.

So, where does this lead? Once again, the question must be put: what are the goals of the poverty industry? If poverty is an inherent distributional element of the structure of capitalism, what role can there be for any poverty industry? What are the aims? Moreover, if poverty is defined as inequality and its eradication is the ultimate goal, it is imperative to take account of Colenutt's (1970) words:

> A war on poverty taken seriously is a call for a new distribution of all resources, including power; this is a call for revolution.

GOVERNMENT INTERVENTION AND THE GOALS OF THE POVERTY INDUSTRY

The degree to which one believes the state to have any valid role in tackling poverty depends upon one's perspective on the structural role of the state within capitalism. Political stances embrace the "right-wing", which might view poverty as the consequence of the low marginal productivity labour and the poor as being undeserving, the victims of their own self-inflicted excesses, of their laziness, stupidity and fecklessness. Intervention may therefore be considered to distort the free labour market and a strategy of benign neglect with social control through police and social workers might represent a preferred policy.

A more "liberal" tradition recognises that society cannot escape blame for failing to provide opportunities sufficient to enable the poor to acquire the necessary attributes (market capacity) to achieve success. Such an approach is likely to stress the role of education and training as tactics in the battle against personal deprivation by providing people with greater capacity to gain more remunerative employment. Liberals (including the "soft-left") seem to possess a considerable faith in the power of the state to transform society, thereby neglecting the state's fundamental role as the guarantor of the existing social relationships of capitalism.

Hutchinson (1969) showed that educational achievement is an important determinant of employment status in Ireland and that marked differences exist between the social classes in their participation rates in higher education. In their excellent analysis of Irish social structure, Breen et al. (1990) demonstrated how, following Independence, there emerged "a class structure in which advantage was allocated increasingly on the basis of educational credentials and less through family property" (Breen et al., 1990, 5). Nevertheless, "the families that enjoyed privileged positions in the old class structure secured comparable positions in the new one while those families at the bottom of the old class hierarchy have, if anything, drifted downward into a new underclass dependent on state income maintenance for their livelihood" (Breen et al., 1990, 17). Thus, in a system where opportunities for upward mobility are highly restricted and where "the inequalities in mobility in Ireland are far more to the disadvantage of men from working class backgrounds than is the case in England and Wales" (Breen et al., 1990, 64), the

> children of middle class parents continue to enjoy a substantial advantage in access to the more privileged occupations. But the corresponding disadvantage of young people from working class origins is increasingly expressed in the risk of becoming unemployed rather than of remaining immobile within the working class (Breen et al., 1990, 67).

One could, therefore, with considerable justification, extend to Ireland Halsey's (1973) observation that the British educational system has remained predominantly an avenue for the stable transmission of status from one generation to another. Moreover, education might merely create higher hurdles for entry into low grade employment. Despite the intervening de-skilling which has resulted from greatly changed office-related technology, jobs into which one could enter straight from school in the late 1960s seem now to be recruiting graduates.

Apart from imparting the idea essential to the disciplining of a future wage-labour force that one should quite "naturally" be obliged to spend large proportions of one's life in locations where one has no desire to be, undertaking tasks which one has no desire to do, formal education also plays the fundamental role of legitimating income inequality by casting the blame for low income on the individual's own under-achievement at school. The limited role of education as a tool for tackling poverty is highlighted by the findings of the Fifth Main Report of the Australian Commission of Inquiry into Poverty and Education (1976) (see Appendix), offering little comfort to those devoted to a gradualist approach to the solution of poverty through education in a society which remains highly stratified.

Radicals, who ascribe to the structure of society the phenomenon of economic deprivation, are more likely to see the solution to poverty in a more fundamental redistribution of income and wealth. However, the "revolutionary left" might equally oppose interventions *via* a welfare state which buys off the assertiveness of working-class wage labourers and suppresses their combativity with modern equivalents of bread and circuses, mindful of the dictum that those to whom one addresses one's demand are those whom one regards as masters of the situation and that "demands-on-the-state reformism" tends to engender clientelism and dependency.

So, how valuable has welfare statism been as an implement in the redistribution of well-being? Research carried out by Meade (1964) into the effects of the welfare state in the UK demonstrated that it served to redistribute income within classes rather than between them, and between age groups within specified classes. The middle-class is more likely to be aware of and to take up the services available, ranging from higher education to screening for cancer. Ireland is somewhat different as its welfare state has been less comprehensive, the middle-class being obliged to fend for itself to a greater extent.

HIDDEN MECHANISMS OF REAL INCOME REDISTRIBUTION

Most would agree that the interventions of the state should at very least be neutral with respect to income, and preferably progressive, redistributing resources from the wealthy to those who are deprived. However, during the late 1970s and in the 1980s, geographers and other social scientists increasingly studied the way in which public goods (health services, personal social services, educational resources etc.) were distributed. Disconcertingly, it was often found, especially in urban areas, that policies which were being pursued by the state and its agencies had unforeseen negative consequences. While one might expect private services, such as private health care in the United States, to favour areas with wealthier residents, a growing body of research began to show clearly that publicly financed services, such as the quality or location of certain aspects of the National Health Service in Britain or the quality of schooling within publicly funded education systems, also favoured the better-off and that public policies frequently possessed highly income-regressive consequences, redistributing real well-being from poor to rich (see Cox, 1973, 1979; Kirby, 1982; Kirby, Pinch and Knox, 1982; Knox, 1978, 1979, 1982; Knox and Pacione, 1980; Pinch, 1978, 1984, 1985).

Sea views, parks and other desirable elements of the urban environment are fought for in the land market and paid for in the price of access to housing in desirable neighbourhoods. At the other extreme, those with little power can be forced to bear the "costs of proximity" to the negative externalities of urban life: the refuse tip and the motorway interchange. These are what geographers call the "hidden mechanisms of real income redistribution" (see Harvey, 1973). We have devised very sophisticated techniques to legitimise these redistributions of well-being. For example, in the inquiry into the siting of London's third airport, sophisticated cost-benefit analyses were used to favour a location which created a noise zone over more populous poorer residential areas rather than over a wealthier area, on the grounds that property values would be adversely affected to a greater extent in the wealthier area!

Because of their powerlessness, the poor can be forced to bear disproportionately the costs of change in the urban environment. For example, although market-related factors constitute a major cause of inner-city decline, the de-industrialisation of inner Dublin was also promoted during the 1970s by a planning authority intent on tidying up the land-use geography of the city, encouraging firms to move out to purpose-built industrial estates at the periphery, thereby ridding the central city of those functions which it deemed to constitute "incompatible land uses". Many industrial functions, offering appropriate employment to an inner-city workforce, found it increasingly difficult to continue operations within the central area. Expansion onto adjacent sites would not only have proved very costly but might well have been precluded by zoning regulations. A survey of 514 firms which had located on industrial estates between 1965 and 1974 revealed that more than a quarter had relocated from the inner city (Dublin Corporation Planning Department, 1975). The Corporation further estimated that this relocation of plants had resulted in a loss of 4,500 inner-city manufacturing jobs during the period. During the recession of the 1970s, the long established industries of the inner-city were particularly badly affected and the Industrial Development Authority estimated that during the late 1970s approximately 2,000 inner-city manufacturing jobs were being lost annually as a consequence of relocation, shrinkage and closure.

Figure 9.1 maps the changes in city centre industrial land-use between 1966 and 1985, revealing the considerable extent of such industrial decline. Between 1966 and 1974, the period of the first modern office development boom, the total area of industrial floorspace in the inner-city declined by over 550,000 sq.m. (5.92 million sq. ft.) or 30 per cent. Meanwhile, office uses increased by nearly 95 per cent, thereby displacing industrial functions as the second most important activity in the inner area after residential uses. Clarke's (1991) investigation of floorspace change in an area lying to the east of Westland Row on the margins of the central business area, reveals that the process of converting industrial land uses to office functions has been very active since the electrification of the coastal rail line (DART) and the enhancement of accessibility in the locality of the railway station at Westland Row. Industrial functions had accounted for around 45 per cent of total floorspace in 1966, but by 1990 this had shrunk to below 5

Figure 9.1: Shrinkage of Industrial Land Uses, 1966–85

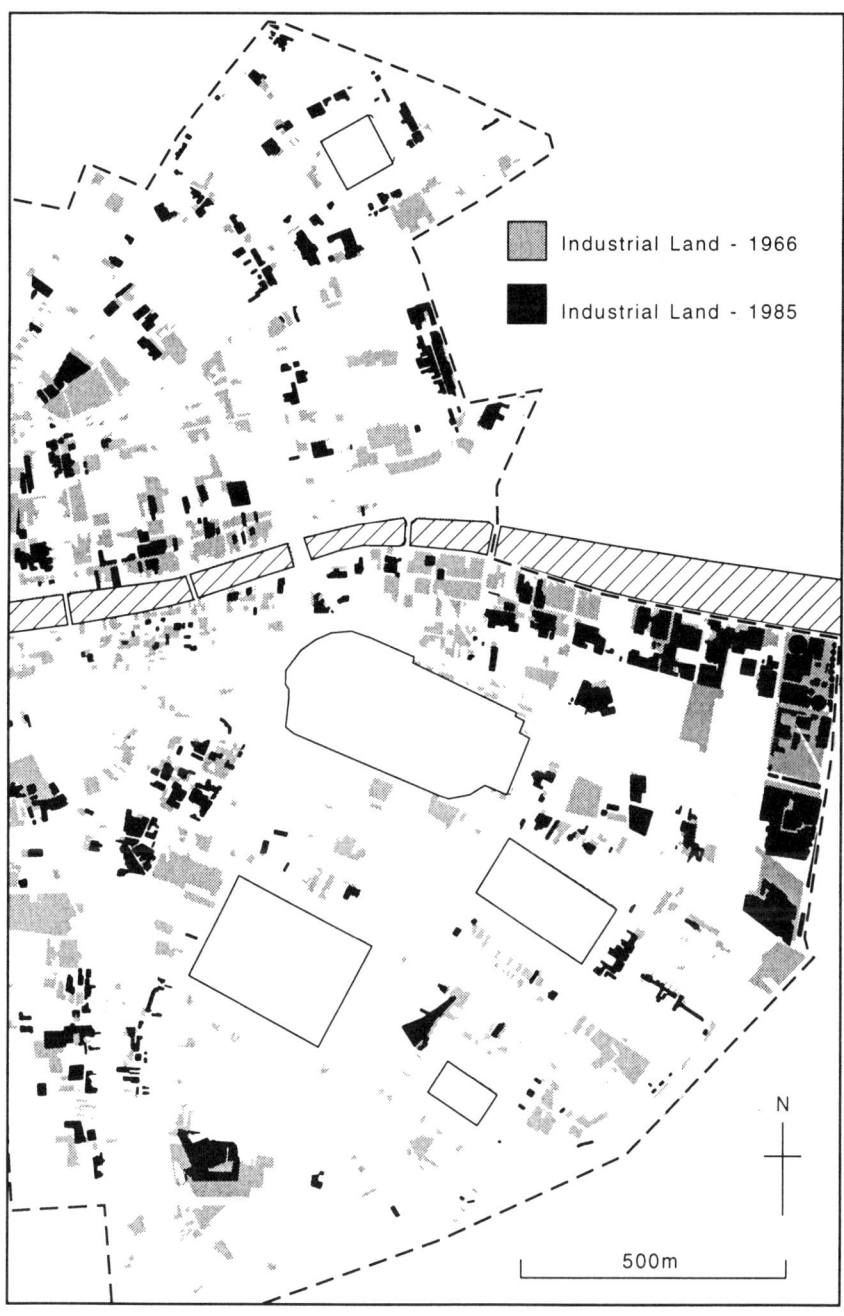

per cent. In contrast, office functions had expanded from 5 per cent to 51 per cent during the intervening years. The negative ramifications for the provision of employment appropriate to the skills of local residents were considerable and constituted a virtual emasculation of the area's industrial economic base. It is a cause of considerable concern that such changes should have resulted, at least in part, from public spending on infrastructural developments and that the possibility of such ramifications was never foreseen by planners.

Simultaneously, public-sector housing policies and government incentives favouring the provision of new housing for owner occupation in the suburbs have resulted in a massive expansion of suburban residential areas since 1950. While the residents of the inner city have been confronted by a dwindling quantity of appropriate employment, they have been provided with a public transport service dedicated primarily to bringing white-collar workers from the dormitory suburbs into their city-centre workplaces, rather than a system geared towards those wishing to travel from the centre to peripheral industrial estates. Residentially immobile and possessing low rates of car ownership, inner-city populations increasingly faced the prospect of long-term unemployment.

With regard to the problems of regenerating inner Dublin, one might well question the relevance to inner-city communities of the Designated Area policies and property-based renewal. These policies initially generated a number of office developments along High Street and along the Liffey's quays (MacLaran, 1996a), followed by a boom in the development of apartments for a population of transient youthful white-collar incomers — a process of gentrification which has been studied closely in recent years (MacLaran *et al.*, 1994, 1995; MacLaran, 1996b; MacLaran and Floyd, 1996). The result has been a welcome reduction in the number of derelict sites and vacant buildings. However, the sale of publicly held land for private sector property development has seriously undermined Dublin Corporation's ability to engage in the development of social housing schemes in the inner city. Moreover, this property boom has also fuelled increases in land values in secondary areas of the inner city. This has created enormous difficulties for community employment programmes in their search for affordable premises, while simultaneously having provided few employment opportunities or affordable accommodation for inner-city residents. Such a policy certainly does not represent a serious commitment to economic regeneration of a type which is relevant to the requirements of inner-city communities. Rather, they may well be seriously detrimental to those who are already deprived.

The current property-based renewal policies being pursued in inner Dublin do have the merit of being relatively cheap, at least in the short term. After all, it is always easier to give away a future government's money through tax reliefs than to commit one's own finances, because one's own party may well soon be out of office! Moreover, property-based renewal possesses highly visible effects, embodied in brand new buildings. Regardless of its irrelevance to and possible detrimental effects upon inner-city residents, property-based renewal gives the impression that something is actually being done, that the state really does care for the wellbeing of the inner-city community. Additionally, the transformation of the inner city into a residential location acceptable to young white-collar workers serves to dilute inner-city social problems. By importing substantial contingents of the non-poor, statistical measures of deprivation recorded on an area basis may be seen to improve. "Deprived areas" become less and less poor despite nothing having been to tackle the problems of multiple deprivation experienced by indigenous inner-city residents.

AREA-BASED APPROACHES: AN IDEOLOGICAL DIVERSION

It is imperative to learn from the past, if only to avoid repeating costly mistakes. During the 1950s, there arose a widely held belief in Britain that the welfare state had banished poverty to small pockets and that the remaining poor comprised individuals whose personal failings prevented their taking advantage of the growing opportunities of post-war recovery. However, during the early 1960s, Harrington's *The Other America*, Audrey Harvey's case studies of homelessness and television programmes like *Cathy Come Home* rapidly disabused society of these fanciful flights of self-deception. The rediscovery of poverty, educational disadvantage, housing deprivation and failing personal social services led to the introduction of a new battery of policy provisions. Originating in the U.S.A., area-based positive discrimination was to become a central plank of government intervention. Programmes were devised which would allocate special resources to those living in designated localities (deprived areas) in order to help fulfil their distinctive needs. Demands have similarly been made for the north inner city of Dublin to be given special provisions, including the creation of an Educational Priority Area (Breathnach, 1976).

Four assumptions underlay such approaches:

1. That there existed small identifiable areas where a high proportion of the residential population was deprived.

2. That the city's deprived were concentrated into these areas.

3. That deprivation could be alleviated by action taken within these areas.

4. That concentrating the available limited resources within a small target locality would generate a neighbourhood positive multiplier effect which would alleviate deprivation more effectively than by distributing the resources more widely.

The late 1960s and early 1970s therefore witnessed a battery of area-based programmes: General Improvement Areas, Educational Priority Areas, Community Development Projects, the Urban Programme, Housing Action Areas, Comprehensive Community Programmmes, Inner-City Partnerships, plus a range of Inner Area Studies. More recently, the Enterprise Zones of the 1980's Thatcherite era sought to strengthen inner-city economies through the relaxation of planning controls together with the inducement of financial incentives such as relief from commercial rates. Their results were varied. It has been maintained that, in the London Docklands, the policies contributed to the displacement of existing low-grade employment, such as road haulage and scrap dealing, which were unsightly and detracted from the environmental quality appropriate to the new image of docklands as a business location, which the London Docklands Development Corporation was charged to promote. However, such businesses had provided employment which matched the skills of the indigenous community, unlike the incoming office jobs. The policies also led to the influx of new young middle-class professional residents, thereby generating a process of gentrification which resulted in antagonism from the indigenous docklands residents.

In other Enterprise Zones, such as Liverpool, a significant proportion of initial economic development represented employment transfers to the rates-free zones from other parts of the city, or comprised warehousing activities with few employees and low wage jobs. Indeed, Liverpool proved a testing ground for many area-based schemes. Between 1969 and 1974, the city received an Educational Priority Area, an Inner Area Study, an Area Management Scheme, General Improvement Areas, Housing Action Areas, a Community Development Project, a Neighbourhood Project, a number of Integrated Youth Opportunity Projects, 146 Urban Aid Projects and an Enterprise Zone. Needless to state, inner Liverpool remains an area with a high proportion of deprived residents.

Some commentators, such as Gray (1975), have suggested that the adoption of area-based approaches was virtually inevitable. A cursory inspection showed that there were areas of physical decay in cities. Common sense suggested that social deprivation would be similarly concentrated. Moreover, local authority officers tended to view deprivation in area-based terms rather than in terms of class (Stewart, Spencer and Webster, 1976).

However, research undertaken during the 1970s, revealed that social deprivation was far from concentrated. Even in the US, it was found that only half the poor were concentrated in the inner-city, the remainder being thinly distributed throughout middle-class suburbia (Morrill and Wohlenberg, 1971). In the UK, while Educational Priority schools catered for a minority of all the educationally disadvantaged, only a minority of pupils attending such schools were considered to be educationally disadvantaged. Even with respect to slum housing, with its seemingly obvious spatial concentration, the suitability of area-based Housing Action Areas began to be questioned (Graham, 1974). Indeed, from research undertaken into levels of living in Dundee (MacLaran, 1974), so widespread geographically was the problem of multiple deprivation that MacLaran (1981, 65) was led to opine that the "only type of area-based policy which is likely to be cost-effective would be one of discrimination against the privileged who were concentrated geographically to a greater degree than the deprived". From a purely cost-effectiveness/efficiency criterion, the value of area-based schemes was therefore increasingly challenged.

Among the sharpest critics of area-based schemes were those working in the Community Development Projects (CDPs), action-research teams of community workers and academics working for neighbourhood rejuvenation through community development. The participants of these government-funded schemes became highly critical of their role and increasingly came to view deprivation in structural terms:

> The problems of urban policy with which they were confronted were the consequence of fundamental inequalities in the economic and political system (CDP-Inter Project Editorial Team, 1977a, p. 5).

They recognised that the areas in which they were operating were actually important to the market economy, functioning as a pool of expendable labour power, to be used in times of boom and discarded during slumps. Furthermore, they explained the existence of such areas only in relation to the advance of other areas, capital moving freely to more profitable locations. They therefore viewed their decline as a consequence of uneven development, an inherent feature of the capitalist system which becomes manifested at international, regional and intra-urban scales. The areas in which they were working were typified by the withdrawal of capital associated with the restructuring of industrial production, frequently consequent upon the concentration of capital through mergers. It was therefore possible to understand local conditions only by comprehending how capital makes use of geographical space. In the context of a lack of control over capital, local intervention seemed to have little relevance.

More generally, the CDPs attacked the whole basis for interventions which were grounded in the neighbourhood. They claimed that by pleading for extra resources for districts with "special needs", the structural problems inherent in capitalism become transformed into the problems of particular localities. Drawing fixed geographical boundaries to the "action areas" effectively distinguishes "them" (the deprived) from "us" (the remainder of society).

> This idea of poverty affecting only small groups in marginal areas is a powerful one for it immediately reduces the scale of the problem. It also carries the implication that those who live outside these areas share no common interests or problems with the deprived within. The working class are effectively split into two and the scene is set for convincing those within that their problems have nothing to do with wider economic and political processes. . . . Once poverty and exploitation have been defined as marginal it follows logically that only minor adjustments are needed to make it go away (CDP-Inter Project Editorial Team, 1977a, pp. 53-54).

Thus, there is no need to impugn the structure of rewards in society at large nor to question either the underlying structural causes or the primacy of capitalist interests. This the CDPs considered to be diversionary, emphasising that area-based policies run the risk of mistaking symptoms for causes.

> It is clear that there are similarities between the way in which the urban problem is being discovered, defined and tackled now, and the way the regional problem was taken up during and after the depression. Both are ways of defining particular problems of capital as problems of certain spatial areas. . . . The importance of this technique is that it diverts attention from the way in which the problems that appear in particular places are really particular manifestations of general problems — problems of the way the economic system operates (CDP-Inter Project Editorial Team, 1977b, p. 55).

The presence of stark contrasts in the conditions prevailing in different sub-areas of our cities does not imply that the causes are peculiarly "urban" or even "spatial". It is important to recognise that within the urban arena, one's place of residence is largely a product of the way in which the housing system operates. It sifts and sorts the population according to its economic capacity (the ability to negotiate barriers to entry) in relation to the cost of housing (the cost of ownership, of private and public renting) and according to each households' needs for accommodation space (broadly, its stage in the family cycle). Urban space is therefore not independent of social structure: the social patterning of the urban environment represents a secondary feature of social structuration produced by the way in which the housing system operates, given the inequalities present in people's market capacities.

CONSIDERATIONS

Social scientists and policy makers ought fully to recognise that spatial factors can constitute important contributors to variations in social well-being, even at the intra-urban scale. Such factors have been briefly examined above. Thus, there may well be sound reasons for place-specific policies. Area-based housing and environmental renewal under the Glasgow Eastern Area Renewal project has been very successful in improving the dwelling stock and the environmental conditions in a locality where social infrastructure was sorely in need of improvement. Nevertheless, it has far from eradicated poverty. Secondly, access to employment of an appropriate type, the local availability of urban facilities and social support services are vitally significant to the well-being of individuals and may constitute a justification for engaging in area-based approaches. Additionally, it should be ensured that those who are already deprived are not additionally burdened by the fall-out from public policies, nor should they be obliged to bear the costs of proximity to negative externalities within the urban environment — the "hidden mechanisms" of income redistribution.

However, beyond these very limited goals, it is imperative to recognise that area-based policies have little meaningful role as an implement in tackling structural poverty and that such approaches possess inherent ideological dangers. One could argue cynically, but quite logically, that the aim of spatial approaches to poverty might be the production of a geographically even spread or a spatially random distribution of deprivation. This would indeed eradicate the geography of poverty, but it would hardly constitute a solution to the problem. Unfortunately, the re-emergence of area-based policies as vehicles for tackling deprivation tends to confirm Karl Marx's (1852) poignant observation that history has a tendency to repeat itself, the first time as tragedy, the second as farce.

NOTES

1. It is imperative to recognise that objects do not possess an intrinsic property of belonging; no inherent "your-ness" or "my-ness". "Property right" describes the social relationship which exists between individuals and things. In slave societies, such property relations also exist between people and other people, who become the object of property-right. The neolithic revolution propelled the extension of the idea of private property right from simple use-values (clothes, simple hunting implements) towards the privatised ownership of the productive forces of society which had hitherto been communally controlled. The numerically few owners then required a force to protect their claims of ownership from the many, thus bringing into existence the first political states and signalling the declining role of the tribal unit founded upon ties of kinship. "Capital" is ultimately nothing more than a relationship; a claim to control over the means of production which is based upon the concept of "ownership right". Ultimately, capital is a relationship, founded upon an idea which exists only in cerebral space.

REFERENCES

Ambrose, P. and Colenutt, R. (1975) *The Property Machine*, Penguin, Harmondsworth.

Atkinson, A. B., Rainwater, L. and Smeeding, T.M. (1995) *Income Distribution in OECD Countries*, Organsation for Economic Co-Operation and Development, Paris.

Breathnach, A. (1976) "Towards the identification of Educational Priority Areas in Dublin", *Economic and Social Review*, 7 (4), 367-382.

Breen, R., Hannan, D.F., Rottman, D.B. and Whelan, C.T. (1990) *Understanding Contemporary Ireland*, Gill and MacMillan, Dublin.

Byrne, S. (1989) *Wealth and the Wealthy in Ireland, a Review of the Available Evidence*, Combat Poverty Agency, Dublin.

Callan, T., Nolan, B., Whelan, B.J., Whelan, C.T. and Williams, J (1996) *Poverty in the 1990s*, Oak Tree Press, Dublin.

Chesher, A. and McMahon, P. (1976) "The distribution of personal wealth in Ireland — the evidence re-examined", *Economic and Social Review*, 8 (1).

Clarke, N.O.R. (1991) *The Social Consequences of Office Development within the Westland Row Area*, unpublished B.A. Mod. dissertation, Department of Geography, Trinity College, Dublin.

Collenutt, R. (1970) "Poverty and inequality in American cities", *Antipode*, 2(2), 55-60.

CDP-Inter Project Editorial Team (1977a) *Gilding the Ghetto: the State and the Poverty Experiments*, Community Development Projects, Nottingham.

CDP-Inter Project Editorial Team (1977b) *The Costs of Industrial Change*, Community Development Projects, London.

Cox, K.R. (1973) *Conflict, Power and Politics in the City*, McGraw Hill, New York.

Cox, K.R. (1979) *Location and Public Problems*, Basil Blackwell, Oxford.

Dublin Corporation Planning Department (1975) *Land Use in the Inner City Area*, Working Paper no. 9, Planning Department, Dublin Corporation.

Eurostat (1997) *Statistics in Focus: Population and Social Conditions*, Eurostat, Brussels.

Graham, P. (1974) "Are Action Areas a mistake?" *Municipal Public Service Journal*, 5 July, 809-811.

Gray, J. (1975) "Positive discrimination in education, a review of the British experience", *Policy and Politics*, 4, 85-110.

Halsey, R. (1972) *Educational Priority*, HMSO, London

Harrington, M. (1966) *The Other America*, Penguin, Baltimore.

Harrison, A.J. and Nolan, S. (1975) "The distribution of personal wealth in Ireland — a comment", *Economic and Social Review*, 7(1).

Harvey, A. (1960) *Casualties of the Welfare State*, Fabian Tract 321, London.

Harvey, D. (1973) *Social Justice and the City*, Edward Arnold, London.

Honohan, P. and Nolan, B. (1993) *The Financial Assets of Households in Ireland*, Research Paper 162, Economic and Social Research Institute, Dublin.

Hutchinson, B. (1969) *Social Status and Inter-Generational Social Mobility in Dublin*, Economic and Social Research Institute, Paper No. 48, Dublin.

Kirby, A. (1982) *The Politics of Location*, Methuen, London.

Kirby, A., Knox, P. and Pinch, S. (1984) *Public Service Provision and Urban Development*, Croom Helm, Beckenham.

Knox, P.L. (1978) "The intra-urban ecology of primary medical care: patterns of accessibility and their policy implications", *Environment and Planning*, A10, 415-435.

Knox, P.L. (1979) "Medical deprivation, area deprivation and public policy", *Social Science and Medicine*, 13D, 111-121.

Knox, P.L. (1979) "The accessibility of primary care to urban patients: a geographical analysis", *Journal of the Royal College of General Practitioners*, 29, 160-168.

Knox, P.L. (1980) "Urban deprivation and health care provision", *Medicine in Society*, 5, 54-59.

Knox, P.L. (1982) "Residential structure, facility location and patterns of accessibility", in Cox, K.R. and Johnston, R.J. (eds.) *Conflict, Politics and the Urban Scene*, Longman, London.

Knox, P.L. and MacLaran, A. (1978) "Values and perceptions in descriptive approaches to urban social geography", in Herbert D.T. and Johnston R.J. (eds.) *Geography and the Urban Environment*, Vol. 1, Wiley, Chichester.

Knox, P.L. and Pacione, M. (1980) "Locational behaviour, place preference and the inverse care law in the distribution of primary medical care", *Geoforum*, 11, 43-55.

Lyons, P.M. (1975) "Estate duty wealth estimates and the mortality multiplier", *Economic and Social Review*, 6(3).

MacLaran, A. (1974) *Spatial variations in levels of living in Dundee*, unpublished Ph.D. thesis, University of Dundee.

MacLaran, A. (1981) "Area-based positive discrimination and the distribution of well-being", *Transactions, Institute of British Geographers*, 6 (1), 53-67.

MacLaran, A. (1993) *Dublin: the Shaping of a Capital*, Belhaven, London, 242 pp.

MacLaran, A. (1996a) "Office development in Dublin and the tax incentive areas", *Irish Geography*, 29(2), 49-54.

MacLaran, A. (1996b) "Private sector residential development in central Dublin", in Drudy, P.J. and MacLaran, A. (eds.) *Dublin: Economic and Social Trends — Volume 2*, Centre for Urban and Regional Studies, Trinity College, Dublin, 20-42.

MacLaran, A. and Floyd, D. (1996) *Recent Residential Developments in Central Dublin*, Centre for Urban and Regional Studies, Trinity College, Dublin, 137pp.

MacLaran, A., MacLaran, M. and Williams, B. (1994) *Residential Development as an Engine for Inner-City Renewal in Dublin: Commentary and Statistical Appendix*, Centre for Urban and Regional Studies, Trinity College Dublin, 53 pp.

MacLaran, A., Williams, B. and Emerson, H. (1995) *Residential Development in Central Dublin: a Survey of Current Occupiers*, Centre for Urban and Regional Studies, Trinity College, Dublin, 106 pp.

Marx, K. (1852) *The Eighteenth Brumaire of Louis Napoleon*.

Meade, J. (1964) *Efficiency, Equality and the Ownership of Property*, Allen and Unwin, London.

Morrill and Wohlenberg (1971) *The Geography of Poverty in the United States*, McGraw Hill, New York.

Nolan, B. (1991) *The Wealth of Irish Households*, Combat Poverty Agency, Dublin.

Nolan, B. (1992) *Low Pay in Ireland*, Research Paper 159, Economic and Social Research Institute, Dublin.

Nolan, B., Callan, T., Whelan, C.T. and Williams, J. (1994) *Poverty and Time: Perspectives on the Dynamics of Poverty*, Research Paper 166, Economic and Social Research Institute, Dublin.

Pinch, S. (1979) "Territorial justice in the city: a case study of the social services for the elderly in Greater London", in Herbert, D. T. and Johnston, R. J. (eds) *Social Problems in the City*, Oxford University Press, 201-223.

Pinch, S. (1984) "Inequality in pre-school provision: a geographical perspective", in Kirby, B., Knox, P.L., and Pinch, S. (eds.) *Public Service Provision and Urban Development*, Croom Helm, Beckenham.

Pinch, S. (1985) *Cities and Services: the Geography of Collective Consumption*, Routledge and Kegan Paul, London.

Scott, J. (1994) *Poverty and Wealth: Citizenship, Deprivation and Privilege*, Longman, Harlow.

Stewart, J., Spencer, K. And Webster, B. (1976) *Local Government Approaches to Urban Deprivation*, Home Office Urban Deprivation Unit, Occasional Paper No. 11, London.

APPENDIX: COMMISSION OF INQUIRY INTO POVERTY: POVERTY AND EDUCATION IN AUSTRALIA

Fifth main report, 1976, AGPS, Canberra.
There is widespread belief in our community that, of all the many solutions to poverty, education more than any other provides the way out for poor people:

> Education plays a crucial role in any anti-poverty program. . . . Broadly speaking, we have a situation in Australia today where some social groups like the professional middle classes recognise the importance of the school, know how to use it and can communicate these values to their children. Their children in turn are able to succeed by meeting the requirements of the system. On the other hand social groups who are disadvantaged, either don't see its impor-

tance in life, misunderstand the purpose of education or simply don't have the skills to compete within the system. Of all our social systems, the school and the pre-school centre are potentially the best opportunity for intervening in the "poverty cycle".

Poor people themselves are often convinced that if they had stayed in school, tried harder and obtained better qualifications, things would be different for them now. The following comments were made in a group discussion by families involved in a welfare agency experimental program.

> Wish I could have gone further in school now. I think a lot of people feel that.

> *What do you mean "further"?*

> So that I could have taken a better position in work. Better money. Like, I've got six children.

> Well. I've got the same problem. My trouble was going to school and not staying. I'm sorry now that I never learnt a trade. I could have got further ahead and bettered myself. I'll go the opposite way now and teach my children to go further ahead than what I did. I don't want them to leave off where I did.

In recognising the greater importance placed on formal credentials in our post-industrial society, poor parents, in common with many others, hope that schooling will enable their children to do better in life than they did.

The general rise in aspirations seems closely related to the impact of technology on the structure of work. Because of a more extensive hierarchy of jobs which are linked with paper credentials and shifts in employment policy, the schools have assumed a more influential role in society. Access to good jobs by way of experience in the school of hard knocks has diminished and life chances are increasingly seen to be determined by the length and quality of formal education. Indeed, as already shown, the rewards offered by society to young people who obtain the final school certificate can include access to a range of worthwhile careers which, in turn, can mean escape from poverty. The fact that some adolescents from poor families have achieved this goal is often used to support the belief that if all poor students were encouraged to stay on longer at school and, by special efforts, were assisted to pass exams and succeed academically, the problem of poverty would be largely overcome.

But the evidence in our Report raises basic queries about the validity of such beliefs on the part of our society. Our investigation of the outcomes of schooling has revealed that success in school and in the competition for rewarding careers is largely determined by such factors as social class, ethnic background and geographic location. The structural inequalities in our society are nowhere more evident than in our school systems. Far from being a way out for poor people, schools act as a sorting, streaming mechanism helping to maintain the existing distribution of status and power. The belief that schools provide the best possible potential for intervention in the poverty cycle is in conflict with the actual function schools carry out in ranking students on the basis of their congruence with school values and consequent academic success. However diffident many teachers may feel about the arbitrary imposition of this artificial process, they continue to conduct the daily routines and exercises by which students compete against one another for prized and privileged places in the workforce.

10

POOR PEOPLE OR POOR PLACE? URBAN DEPRIVATION IN SOUTHILL EAST, LIMERICK CITY

Des McCafferty
Mary Immaculate College, University of Limerick

INTRODUCTION

In its overview statement on *Poverty, Social Exclusion and Inequality in Ireland*, the Inter-Departmental Policy Committee on the National Anti-Poverty Strategy identifies concentrations of socio-economic disadvantage in two general types of urban setting (Government of Ireland, 1995). These are: (a) inner-city areas that have experienced economic and physical decline and restructuring; and (b) large local authority housing estates on the outskirts of urban centres. While the distinction between these two types may be overstated, in particular in terms of the causal factors and processes involved, both types of area are found in Limerick, the third largest urban centre in the country. Research based on the 1991 census of population has demonstrated the existence of a large number of areas characterised by relative deprivation within the County Borough of Limerick. Using a composite index of deprivation derived from thirteen surrogate measures of poverty, sixteen of the 37 wards of Limerick CB have been classified into the most disadvantaged decile (10 per cent) of census districts in national terms (GAMMA, 1995). These sixteen wards account for some 48 per cent of the population of the County Borough. At the other end of the spectrum, six wards, containing 20 per cent of the population, rank in the least disadvantaged decile in national terms. Limerick is clearly a highly polarised city in socio-economic terms. Moreover this polarisation has a strong geographical dimension, with wards in the most deprived quintile (i.e. with decile scores of 9 or 10 — see Figure 10.1) located in a classical sectoral pattern running north-west to south-east through the city centre. Despite recent strong growth in investment in the local economy, disparities in levels of socio-economic well-being have not diminished; indeed they may have increased (PAUL, 1994).

This paper focuses on social, economic and physical (environmental) aspects of poverty in part of the Southill area of Limerick, an area of public housing located on the southern edge of the city. Southill falls within the most disadvantaged decile referred to above, and was one of the four communities in Limerick city targeted under the EU Poverty 3 programme which ran from 1989 until 1994. The area was developed at a time of major housing shortage in the late 1960s and early 1970s in what was then the largest housing development ever undertaken by Limerick Corporation. While most of the initial tenants were established residents of older areas of the city, who were either living in overcrowded accommodation or displaced by redevelopment, a significant proportion consisted of recent migrants from rural areas who were drawn to the city by the expansion of job opportunities, especially in the manufacturing sector. Southill, which is usually defined at parish level, comprises the four housing estates of Carew Park, Keyes Park, Kincora Park and O'Malley Park, which together contain approximately 1150 households (Figure 10.2).

Because the Southill estates are all broadly similar in socio-economic character and were built at about the same time[1], the area has a strong identity within the city. This identity is reinforced by the fact that Southill is delimited by clearly defined boundaries on all sides: to the south and west by undeveloped land or land zoned for recreational use, and to the north and east by industrial land uses and major regional roads. For the purposes of this paper, attention is confined to the eastern part of Southill comprising O'Malley Park and Keyes Park which together constitute the Galvone B census ward. These two areas, and in particular O'Malley Park which is by far the largest of the four estates, constitute the core of the Southill area, containing roughly 66 per cent of its total population. While the sub-area accounts for just 5 per cent of the population of the County Borough

Figure 10.1: Ward Deprivation Scores, 1991, Limerick County Borough

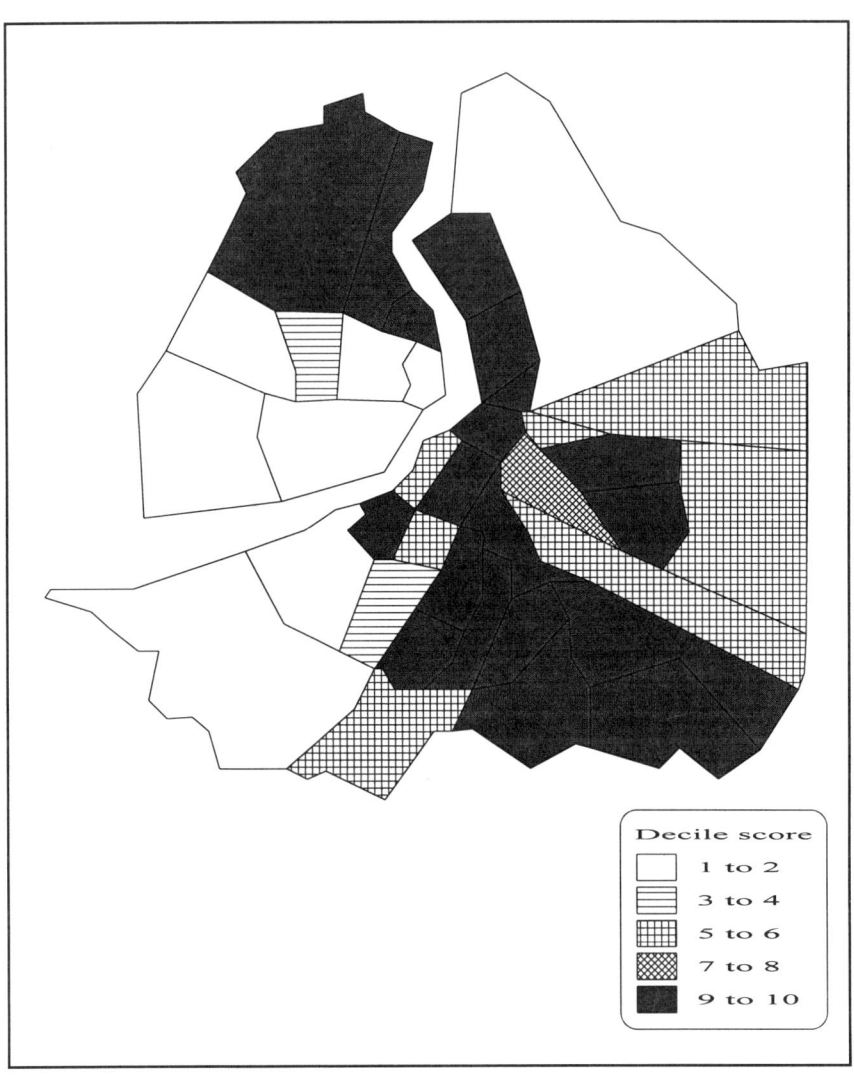

Figure 10.2: Location Map

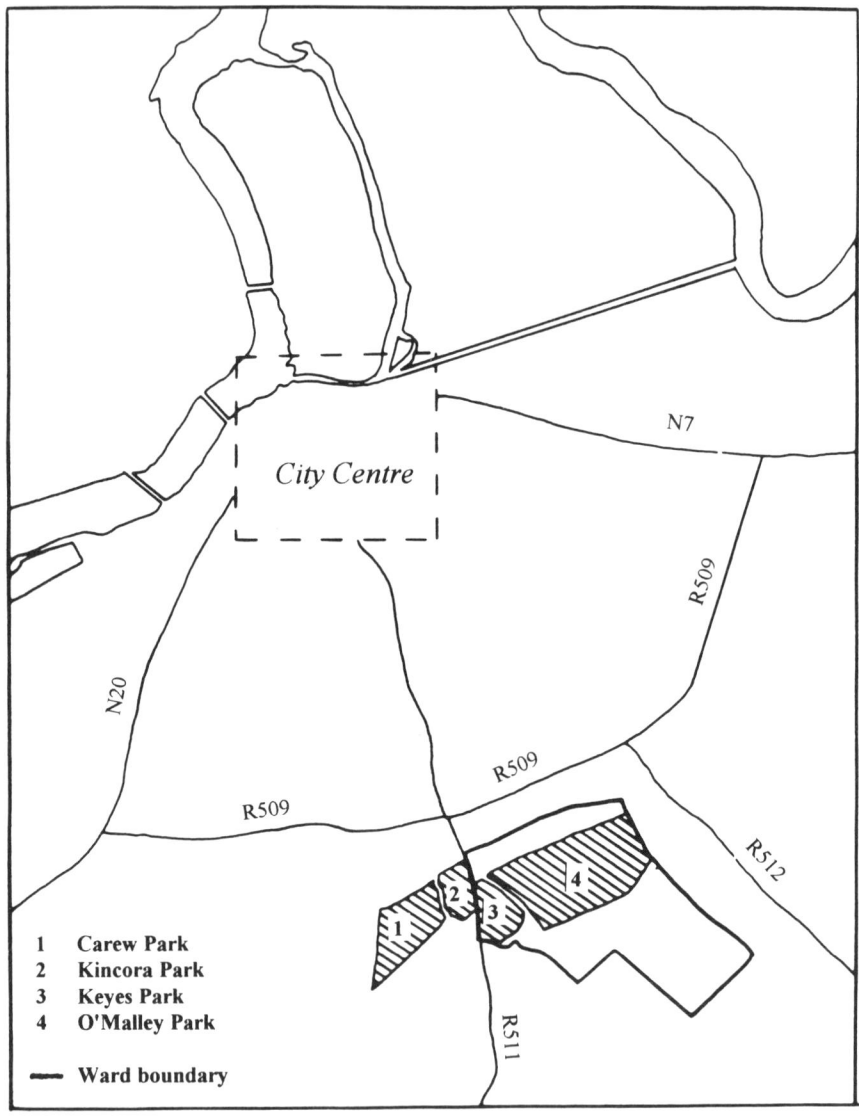

as a whole, it exemplifies many of the social, economic and environmental problems that are common to deprived communities in Limerick and other Irish towns and cities, in particular those in large peripheral public housing estates. The sub-area will be referred to throughout the paper as Southill East.

The remainder of the paper is divided into three sections as follows. The first section looks at a number of aspects of the demographic, social and economic structure of the area, focusing in particular on those attributes which have been shown in national studies to characterise households which are at a high risk of poverty. The next section switches the focus to what Smith (1977) refers to as "place poverty". It examines key aspects of environmental quality in the area (including housing), and argues that deficiencies in respect of the physical environment have played a central role in the social and economic deterioration of the area. The final section concludes the paper with a brief discussion of the need for policy initiatives, under the rubric of area-based initiatives in particular, to address the problems of deprived peripheral housing estates.

ASPECTS OF THE DEMOGRAPHIC, SOCIAL AND ECONOMIC STRUCTURE OF SOUTHILL EAST

This section of the paper provides a profile of the population of Southill East based on the small area population statistics from the 1991 census of population, which furnishes information on a wide range of demographic, social and economic variables[2]. The focus is on characteristics which are specially relevant to an understanding of the nature of disadvantage in the area, in particular on those attributes which have been shown in national studies based on the 1987 ESRI survey of income distribution to be associated with income poverty, or with income poverty and material deprivation (see for example Callan and Nolan, 1988; Callan *et. al.* 1989 and 1993). While the significance of particular attributes depends to some extent on the measure of poverty used, certain household types have been consistently identified as being at a high risk of poverty in the sense that relatively high proportions of such types fall below the various threshold income levels used to establish poverty lines. Among the subset of these for which small area enumerations are available from the census of population are: households with more than four children; households with young female heads; single adult households with children; and households headed by an unemployed person. By looking at the rate of occurrence of these household types in the area, and more generally at the attributes on which they are based (large family sizes, lone parenthood, unemployment), it is possible to get a preliminary picture of the extent of poverty in Southill East.

Population, Household and Family Structures

In 1991 the population of Southill East stood at 2,748 persons. The age structure of the population is illustrated in Table 10.1, which shows that, relative to the city as a whole,[3] the area has a youthful population profile with a preponderance of those aged under 25 years of age, and relatively few aged over 55 years. The cohort aged under 15 years accounted for over one thousand persons — more than one-third of the entire population — and this is reflected in a young dependency ratio[4] of 61.6, over one and a half times that for the city as a whole (38.8). While some maturing of the age profile has occurred in the 25 years since the development of the area — the young dependency ratio in 1971 stood at 106.7 — this has been less marked than might have been expected, especially in view of the general decrease in fertility rates. The persistence of the youthful population profile is related to the pattern of population movements from the area. These movements are central to the problems of the Southill East area and are explored in more detail in the next section.

Table 10.1: Population Distribution by Age Group

	Southill East		Limerick City
Age Group	*Number*	*%*	*%*
0–14 years	1,030	37.5	25.4
15–24 years	647	23.5	20.7
25–39 years	509	18.5	21.4
40–54 years	416	15.1	15.9
55 years and over	146	5.3	16.6

The average household size and average family size in Southill East are above the respective city averages, and 21 per cent of families have four or more children as compared to 14 per cent in Limerick city. In keeping with the relatively youthful population structure, families tend to be at the earlier stages in the family cycle (Table 10.2). One-third of all families are at the pre-school or early school stages of the cycle while in over half of all families with children the eldest child is aged under 15 years of age.

Table 10.2: Stage in Family Cycle

	Percentage of Families	
Stage in Family Cycle	Southill East	Limerick City
Pre-family	3.6	6.3
Pre-school	17.4	11.3
Early school	15.9	12.0
Pre-adolescent	15.4	13.3
Adolescent	19.8	18.6
Adult	24.4	26.9
Empty nest or retired	3.4	11.7

A further aspect of the youthful demographic character of the area is the fact that a large proportion of households are headed by young people. Some 29 per cent of heads of households are aged under 30 years of age as compared to 16 per cent in the city as a whole. Significantly, 66 per cent of young heads of household, so defined, are female, and households with female heads aged under 35 years constitute 19 per cent of all households in Southill East, as compared to 8 per cent in Limerick city as a whole. The age and sex profile of heads of households suggests a high rate of household formation in the area driven by a relatively high rate of lone parent families. In fact, lone parent families account for 33 per cent of all families, and for 36 per cent of families with children in Southill East, levels which are roughly twice the respective rates for the city as a whole (16 per cent and 20 per cent). More significantly, lone parent young families, defined as those with all children aged under 15 years, constitute 45 per cent of all young families in the area, roughly three times the city rate (15 per cent)[5]. Overwhelmingly, these families have female heads (Table 10.3).

Table 10.3: Lone Parent Young Families (all children under 15 years)

	Percentage of All Young Families	
	Southill East	Limerick City
Lone father	3.03	1.40
Lone mother	41.75	13.18
Total	44.78	14.58

Labour Force Characteristics

The socio-demographic patterns noted above have implications for the functioning of the labour market in Southill East, most obviously through their effects on the labour supply. Recent research has shown a link between family structures and educational achievement (Gordon, 1996), while educational achievement in turn is widely recognised as a key factor in determining the individual's labour market prospects (see, for example, Breen, 1991; Nolan *et. al.*, 1994). Both the male and female labour force participation rates are above the respective city norms, but this is largely a function of the age structure of the labour force, which is heavily weighted towards the younger age groups. When the age-specific rates are examined it is clear that below average participation rates obtain in all age groups other than the cohort aged 15 to 24 years (Table 10.4). Lower than average rates amongst the population aged 25 years and over are due mainly to relatively high proportions of females engaged in home duties, especially in the age group 25 to 34. This results in a female participation rate for that age group which is just 57 per cent of the city average. The relatively high activity rates in the 15 to 24 years age group are related to a low level of participation in education: 24 per cent of 15 to 24 year olds locally are engaged in full-time study, as compared to a city rate of 50 per cent. This is consistent with generally low levels of educational attainment in the area, as reflected in the fact that 57 per cent of the population aged over 15 years ceased education aged 15 or younger, a rate of early school leaving that is almost twice that of the city (30 per cent).

Table 10.4: Age- and Sex-Specific Rates of Labour Force Participation

	Southill East		Limerick City	
Age group	*Males*	*Females*	*Males*	*Females*
15–24	73.3	55.8	46.2	45.5
25–34	92.9	36.3	94.3	63.7
35–44	87.2	28.0	93.5	43.2
45–54	78.4	31.2	86.1	36.1
55–64	48.9	16.7	57.9	19.7
65 +	5.0	0.0	8.3	2.6
Total	77.5	39.2	67.2	38.7

Given the low levels of educational attainment in the area, it is not surprising that unemployment is particularly acute in Southill East. In fact, unemployment levels are significantly above city-wide levels for all age groups, and the aggregate unemployment rate is almost two and a half times the overall city rate (Table 10.5). In general, the highest unemployment rates are found amongst the younger age groups, and this together with the higher participation rates of these age groups results in a concentration of unemployment among the younger population: 71 per cent of the unemployed in the area are aged under 35 years. Of these, 35 per cent have never been in employment. In line with the overall trend, the rate of unemployment among heads of household (31 per cent) is close to two and a half times the corresponding city rate.

Table 10.5: Age-Specific Unemployment Rates

Age group	Southill East	Limerick City
15–24	53.3	30.9
25–34	53.7	17.3
35–44	49.4	18.2
45–54	30.2	16.3
55–64	34.4	16.5
65 +	0.0	9.7
Total	48.8	20.5

Limitations of Surrogate Measures of Poverty

The above discussion has highlighted a number of characteristics of households and individuals in Southill East which are relevant to any attempt to assess the level of poverty in the area. In summary, the area has a very youthful population structure, larger and younger families than average, a significant proportion of households with young female heads, a high rate of single parent families, low levels of post-primary educational attainment, and high levels of unemployment, especially youth unemployment. All of these are characteristics known to be associated with higher than average risks of poverty, so that the data presented can be interpreted as evidence that poverty is more concentrated in the area than in the city as a whole. Among the household/family categories considered here, those which are known to have the highest relative risks of poverty are households / families headed by single parents, and those with an unemployed head. Significantly, these are also the categories which show the highest levels of relative concentration in Southill East. The general thrust of these findings is supported by other indicators of poverty such as measures of consumption. For example, car ownership is considerably below the city average with 23.8 persons per car in Southill East as compared to a level of 4.7 in the city (based on 1991 census data). Similarly, 85 per cent of households in the area are without a car, almost twice the corresponding city rate (44 per cent).

However, a number of caveats need also to be borne in mind. First, given that the 1996 census small area statistics are not available at the time of writing, the data presented are obviously somewhat dated, which is quite a serious problem in the light of the high level of population turnover in the area (see next section). Second, there is the problem of the ecological fallacy: it must be remembered that the co-incidence of socio-economic problems at the census ward level tells us nothing about their possible co-incidence at household level, and it is the latter factor which affects both the intensity and pervasiveness of poverty within the area. Third, the approach adopted so far has essentially been a probabilistic rather than a deterministic one: it indicates the *likelihood* of a higher than average proportion of the population living in poverty but does not constitute proof that this is the case. Finally, and perhaps most importantly, a number of those interviewed as part of the present research, including local community representatives and activists, were strongly of the opinion that income poverty and material deprivation were *not* widespread throughout the area, due in large part to the operation of such factors as the black economy in augmenting both income and consumption opportunities for households.

In the light of these considerations it is necessary to adopt an alternative perspective in order to explore the nature and extent of poverty in Southill East. This involves an examination of the characteristics of the physical environment such as housing and estate design. In other words, we can look at characteristics of the place rather than characteristics of the people.

HOUSING AND ENVIRONMENT IN SOUTHILL EAST

While not everyone in Southill East is affected by problems such as unemployment, the bundle of externalities which together constitute the quality of the environment impinges on all residents. For groups such as the unemployed and single parents, whose activity patterns are less likely to include extended periods outside the neighbourhood, the quality of the residential environment is particularly important. However, environmental quality is notoriously difficult to measure as there are many different aspects to it, and those which are considered important by outsiders may not be so regarded by residents (Knox, 1995, pp. 30-36). Even among residents, views on the relative importance of various components of environmental quality are likely to differ according to such personal factors as age, family circumstances and employment status.

Poor access to services has been widely identified as central to the problems of peripheral housing estates in Irish cities (e.g., Department of the Environment, 1996), and in Southill East groups such as the elderly and single parents face considerable constraints on activity patterns as a result of the relative peripherality of the estate. Elsewhere, Robertson (1984) has suggested that such constraints may be especially severe for lone parents in employment. However, accessibility does not appear to be a critical problem in the area in general, and services such as shops (including a sub-post office), health services, a crèche and pre-schools, primary schools, a community college, a community centre and recreational facilities are located either in, or within walking distance of, the estates. Moreover, while levels of car ownership are low, a reasonably frequent bus service operates between the estates and the city centre. Some indication of residents' priorities in relation to environmental issues is provided by a recent study commissioned by an estate management group in the area which was designed to elicit residents' views on various aspects of the local environment (McCafferty, 1994). When asked to indicate which issue (from a pre-determined list) they felt should receive priority in an action programme for their estate, the majority (60 per cent) of respondents said that the condition of housing was the most important issue, with the second highest proportion (21 per cent) prioritising the design and layout of the estate itself. This section begins with an examination of these two aspects of environmental quality, based in part on the results of the 1994 survey, and on the action plan subsequently prepared by Limerick Corporation (Limerick Corporation, 1995).

Housing Design and Maintenance
The estates in Southill East were built by the National Building Agency to a standard design. The housing consists predominantly of two-bedroom dwellings in terraces, with a small number of one-bedroom bungalows. While the small size of housing together with

the larger than average household size gives rise to a relatively high level of persons per room — 0.88 as compared to 0.63 in the CB — this does not appear from the 1994 survey to be a critical issue for most householders (McCafferty, 1994). In contrast however, the situation with regard to the heating of houses was considered to be unsatisfactory or very unsatisfactory by 68 per cent of respondents. In the case of Southill East these problems arise not so much from poor housing design, but from inadequate maintenance. Indeed, general maintenance appears to be the least satisfactory aspect of housing conditions, with 81 per cent of tenants describing the situation as very unsatisfactory. Increasingly too there have been complaints about the condition of houses at the point of re-letting, with some houses reportedly lacking basic fixtures and fittings. A city-wide maintenance survey conducted by Limerick Corporation in 1994 showed that in a large part of the area scores on such items as windows and external doors, fireplaces and roof insulation were significantly below the overall CB average for public rented housing (Limerick Corporation, 1995). While some households have been able to compensate for the inadequacies of the maintenance system by undertaking repairs themselves, this is obviously more difficult for groups such as the elderly and single parents.

Estate Design
The design of the estates in Southill East suffers from a number of flaws, in functional as well as in aesthetic terms. Both O'Malley Park and Keyes Park are based on the Radburn layout whereby the traditional relationship between houses and access roads is inverted: houses face onto open green areas, and vehicular access is via cul-de-sacs or "back courts" at the rear of houses. The central principle of the Radburn design is the separation of vehicular and pedestrian traffic, but while this may have been an important consideration in cities where the growing level of vehicular traffic was a major problem (Relph, 1987, p. 65), the advantages have been less significant in Southill East given that levels of car ownership are relatively low.

The inversion of the usual relationship between houses and the street has had a number of negative consequences. The provision of vehicular access at the rear of houses creates problems of access for visitors to houses, for taxis and for emergency services. Secure parking is also a problem as there is no provision for this within the curtilage of houses. From the outset, the back courts created problems of security and privacy, but the replacement of the original wooden fences which demarcated the back yards of houses with six foot high walls means that the courts cannot be supervised from the houses. This has led to the marked deterioration of the condition of the back courts in a number of areas. Along much of the boundary of O'Malley Park the houses face away from the rest of the estate, a design feature which from the outset was not conducive to community development in an area newly settled by families drawn from diverse areas within and beyond the city. The large size of the latter estate — 601 houses were constructed initially — has had a similar effect, and efforts subsequently to develop individual housing areas and neighbourhood identities have been compromised by the fact that both the houses and housing areas have few distinguishing features, and by the retention of the original 1–601 numbering system.

Apart from these basic design problems the area suffers from too much large-scale open space which is poorly landscaped. The streetscape quality is poor along the loop road which provides the only access to O'Malley Park. The latter is abutted by gable ends of terraces which are not overlooked and lack protection in the form of side gardens or walls. These and other spaces throughout the estate which are deficient in defensible space terms have been subjected to littering, graffiti and vandalism (Limerick Corporation, 1995). The poor condition of the back courts, a lack of planting on the green areas and the poor streetscape all combine to present a rather bleak landscape.

These weaknesses in the estate design were exacerbated from the outset by the imbalanced demography of the area and in particular the extremely high concentration of children noted earlier. Blackwell (1988, p. 157) notes that high child densities especially under conditions of relative poverty cause the local environment to "wear out" more quickly. Page (1996) suggests that high child densities are central not just to problems of environmental deterioration in residential areas, but also to problems of vandalism and minor incivilities, which in turn can serve as catalytic factors in the spiral of neighbourhood decline.

Population Flows and Estate Management

The result, and in many respects the most telling indication, of the problems noted above has been manifested in a high level of out-migration from Southill East which has led to a marked decline of population. After the initial growth of the 1970s, which produced a population peak of 3,704 persons in 1981, the population decreased by some 26 per cent up to 1991. This decrease, which in proportionate terms considerably exceeded that pertaining both to the CB and the city, accelerated in the latter part of the decade (Table 10.6).

Table 10.6: Annual Percentage Rates of Population Change, 1971-1991

	Southill East	Limerick CB	Limerick city
1971–81	2.43	0.61	1.83
1981–86	-2.08	-1.51	0.27
1986–91	-3.80	-1.54	-0.29

A large part of the movement out of the area was due to the very high levels of emigration which were endemic throughout Ireland in the mid- to late 1980s, as well as the trend towards decentralisation of population experienced by all the major cities in this period (see Cawley, 1996). Some of the movement could also have been expected as a result of the maturing of the area in the 1980s. However, while these national and local processes would suggest an out-flow concentrated among the young adult population, migration from the area was in fact more widely spread across the age groups. This is revealed by projecting the 1981 population — disaggregated by age and sex — forward to 1991 and then comparing this with the actual 1991 population. The projected 1991 population is found by allowing for (a) the number of deaths in each age-sex group that would be expected over the period if national age- and sex-specific mortality rates applied, and (b) the amount of net migration that would be expected if the age- and sex-specific net migration rates of the CB as a whole over the same period had applied[6]. The difference between the projected and actual population of each age group reflects what can be termed differential migration i.e., population movements beyond those in line with city wide trends (by focusing on the population aged 10 years and over we can ignore changes due to births during the period).

The results of this exercise (Figure 10.3 and Table 10.7) indicate that the 1991 population of the ward aged 10 years and over was some 858 below what would have been expected on the basis of the 1981 population, if the migration rates of the CB as a whole had applied. While the population aged 30 to 39 years in 1991 was slightly higher than expected, the group aged 40 to 54 experienced a differential out-flow of some 245 persons, and those aged 10 to 19 years contributed 422 persons to the differential out-flow. What this suggests is that there was a considerable movement out of the ward of those in early middle-age — the age group which contains many of the original householders of the area (i.e., those aged 20 to 34 in 1971).

Figure 10.3: Differential Migration, 1981–91

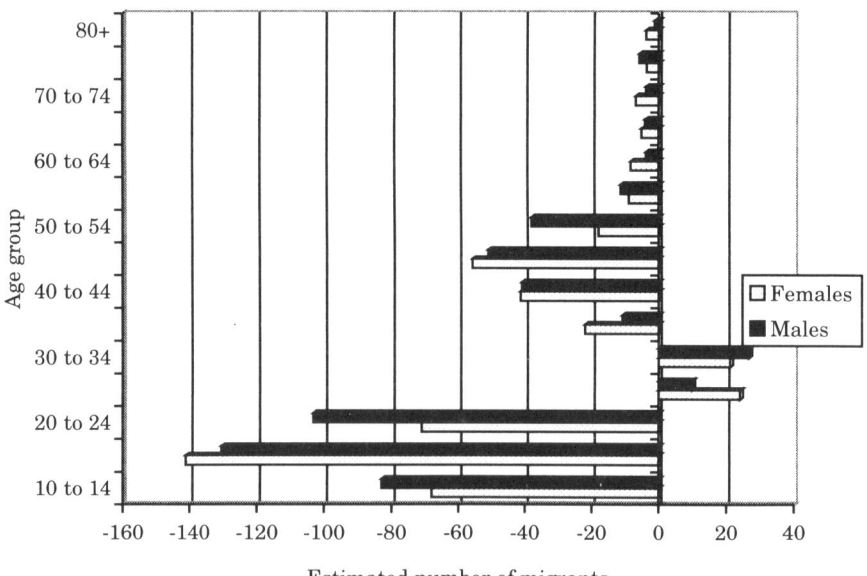

Table 10.7: Age-and Sex-specific Levels of Differential Migration, 1981-91

Age Group, 1991	Males	Females	Total
10 to 19 years	−213.30	−209.17	−422.47
20 to 29 years	−93.36	−46.74	−140.11
30 to 39 years	16.09	−0.81	15.28
40 to 54 years	−129.86	−114.89	−244.75
55 + years	−29.92	−36.47	−66.40
Total	−450.36	−408.09	−858.45

This out-migration of householders was facilitated by a number of factors. Among these is the low level of uptake of the tenant purchase scheme in the area, a facility which has been recognised as "a key factor in the stabilisation of housing estates, in the upkeep of the houses and in promoting community development" (Department of the Environment, 1996, p. 102). Twenty-five years after the building of the estates in Southill East, 63 per cent of the housing stock is still on rent, as compared to just 42 per cent in Limerick CB as a whole.[7] A more important factor however was the surrender grant scheme in operation between 1984 and 1987 which encouraged local authority tenants to purchase private sector housing. The scheme is generally regarded as having had detrimental social consequences for poorer estates where the differential out-flow of higher income households and households in employment served to depress income levels further and increase area unemployment rates (Blackwell, 1988, pp. 188-189). The impact of the scheme in Southill East was particularly severe. In addition to the economic effects noted by Blackwell, the scheme decimated voluntary activity in the area, creating a major crisis for such activities as the local scouts group, the local community games association and the youth club, all of which barely survived.[8]

The differential migration pattern from the area has had a number of adverse consequences. It has served to arrest the normal maturing of the age profile by maintaining relatively high child densities/young dependency ratios. More generally, the exodus of householders has created major difficulties for housing management, by creating a high level of turnover of tenancies and, more seriously, delays between lettings which result in houses remaining vacant for some time. Between 1990 and 1993 inclusive there were 560 official re-lettings in a part of the area which contains 418 houses on rent (Limerick Corporation, 1995). In addition to this there was a considerable (but unknown) number of informal re-lettings not notified to the Corporation. Vacant houses have created major problems for residents as they are subjected to extensive vandalism and are often used for anti-social activities. The fact that these vacancies exist at a time when there are over 600 on the Corporation waiting list for housing is indicative of the low level of demand for housing in the area. In this situation, housing in the area tends to be allocated to those who have relatively low priority on the housing list (single parents living at home are often so classified) or who are most desperate for housing. The result is the increasing concentration in the area of those who are most marginalised in the local society.

In conclusion, there is some evidence to suggest that parts of Southill East are caught in a downward spiral, where movement out of the area has either exacerbated existing problems or created a range of new problems that together have the effect of further undermining demand for housing in the area, and increasing the out-flow of residents. While there are many components of this spiral, there seems little doubt that environmental quality is a central factor. This conclusion is supported, for example, by Pacione's (1982) research, in a markedly similar context in Glasgow, which showed that the poor general environment was the most important single reason for movement from a deprived local authority estate.

CONCLUSIONS: PERIPHERAL POVERTY AND PUBLIC POLICY

This paper has examined a number of aspects of urban deprivation in the Southill East area of Limerick city. While it is obvious that the difficulties of the area cannot be solved simply by pouring concrete or indeed planting trees, it should also be clear that action on the environmental front is essential if the problems of the area are to be tackled successfully. McGregor and McConnachie (1995) identify some of the potential benefits from exploiting the complementarities between the physical and economic regeneration of socially excluded neighbourhoods, especially where physical regeneration is based on housing refurbishment and environmental upgrading. These include the provision of highly visible local jobs for which local residents are likely to possess the requisite skills. The limitation of using physical regeneration as a means of economic revival is, as always, that there can be no guarantee that jobs created will go to local residents. It is for this reason that the benefits of environmental improvement should be seen not so much in terms of economic spin-offs but as a means of breaking the vicious cycle of out-migration and deprivation described above. Moreover, while the economic aspects of deprivation require actions (at least those on the demand side) to be taken at the level of the local labour market area, or at regional or national level, solutions to the environmental problems of the area are most appropriately sought at local level. In fact, the case for an area-based local action programme in Southill East has, arguably, a stronger basis in environmental considerations than in considerations such as the local incidence of the city's unemployment problem.

Issues of estate design and layout (including landscaping) as well as estate and housing management are central to any such programme. On the question of estate management there is some cause for optimism, in particular following the publication of the 1991 Plan for Social Housing (Department of the Environment, 1991) and the subsequent Housing (Miscellaneous Provisions) Act of 1992 which introduced measures to improve the management of local authority housing. An active estate management group in O'Malley Park has successfully sponsored a number of environmental initiatives, and while realistically it will be some time before the group is able to assume the degree of responsibility for estate management which is envisaged in the 1991 plan, a start has been made.

On the question of estate design, the picture is more mixed at present. Following an extensive round of consultations with residents, a comprehensive redesign plan for O'Malley Park costed at £4.65m (net of VAT and professional fees) was drawn up by Limerick Corporation in 1995. However, the plan failed to gain the approval of the Department of the Environment, apparently because it did not include housing measures. A renewed application to the Department in November 1996, which placed the emphasis on housing refurbishment, was more successful, and £680,000 has been allocated for this purpose in 1997, with the possibility of further funding in future years.[9] Limerick Corporation has applied these funds to refurbishment in targeted areas of O'Malley Park, where there would appear to have been a drop in the turnover of tenancies.

Notwithstanding these developments, it remains the case that there is a low priority accorded to general environmental improvements in residential areas at central government level, and a lack of discretionary funding for same at local government level. In this context it is worth noting that the total budget for the original O'Malley Park redesign plan is comparable to that which has been allocated to Limerick Corporation for environmental improvement in the city centre under the Urban and Village Renewal sub-programme of the Local Urban and Rural Development Programme. The latter sub-programme (and indeed the earlier urban renewal initiative introduced in 1986) was introduced in large part because of the perceived need to refurbish highly visible city centre areas in order to generate tourism and economic investment. The danger is that peripheral and therefore less-visible housing estates like those in Southill East, which were built as part of the process of inner-city re-development, might now be neglected as the emphasis in development policy swings back towards the city centres.

NOTES

1. Keyes Park which was begun in 1966 is the oldest, and O'Malley Park which was begun in 1969 is the most recent development.

2. Small area statistics (phase II) from the 1996 census were not available at the time of writing. However, subsequent analysis of these data indicates no significant change with respect to the relative position of the study area on the indicators considered here.

3. Throughout the paper comparisons are made with either the County Borough, or the city as a whole (i.e., County Borough plus environs as defined by the census authorities).

4. The young dependency ratio is defined as the number of persons aged under 15 years per hundred persons aged between 15 and 65.

5. Limerick city itself has a significantly higher rate of occurrence of such families than the national average — 10.7 per cent.

6. Five-year survivorship and net migration rates were used to project first from 1981 to 1986 and then from 1986 to 1991.

7. It should be noted however that there are marked contrasts within the area in the level of tenant purchase, which in Keyes Park is over 62 per cent.

8. Ironically, the recent increase in the level of public housing construction is having similar negative consequences, inducing those on the housing waiting list to "hold out" for the prospect of being allocated a newly built house rather than accept an older house in areas such as Southill East.

9. The Corporation's 1996 submission included as supporting documentation an earlier draft of this paper.

REFERENCES

Blackwell, J. (1988) *A Review of Housing Policy*, National Economic and Social Council, Dublin.

Breen, R. (1991) *Education, Employment and Training in the Youth Labour Market*, Economic and Social Research Institute, Dublin.

Callan, T. and Nolan, B. (1988) "Family Poverty in Ireland. A Survey Based Analysis", in Reynolds, B. and Healy, S.J. (eds.) *Poverty and Family Income Policy*, Conference of Major Religious Superiors, Dublin.

Callan, T., Nolan, B., Whelan, B.J., Hannon, D.F. and Creighton, S. (1989) *Poverty, Income and Welfare in Ireland*, Economic and Social Research Institute, Dublin.

Callan, T., Nolan, B. and Whelan, B.J. (1993) "Resources, Deprivation and the Measurement of Poverty", *Journal of Social Policy*, 22(2), 141-172.

Cawley, M. (1996) "Town Population Change in the Republic of Ireland: The Need for an Urban Policy", *Regional Studies,* 30(1), 85-89.

Department of the Environment (1991) *A Plan for Social Housing*, Stationery Office, Dublin.

Department of the Environment (1996) *Ireland. Habitat II National Report*, Stationery Office, Dublin.

GAMMA (1995) *Limerick City APC Report*, ADM Ltd, Dublin.

Gordon, I. (1996) "Family Structure, Educational Achievement and the Inner City", *Urban Studies*, 33, 407-423.

Government of Ireland (1995) *Poverty, Social Exclusion and Inequality in Ireland. An Overview Statement*, Inter-departmental Policy Committee on the National Anti-Poverty Strategy, Dublin.

Knox, P. (1995) *Urban Social Geography. An Introduction*, 3rd edition, Longman Scientific and Technical, Harlow.

Limerick Corporation (1995) *O'Malley Park Redesign Report*, Limerick Corporation, Limerick.

McCafferty, D. (1994) *Redesign Study for O'Malley Park*, O'Malley Park Estate Management Group, Limerick.

McGregor, A. and McConnachie, M. (1995) "Social Exclusion, Urban Regeneration and Economic Reintegration", *Urban Studies*, 32(10), 1587-1600.

Nolan, B., Callan, C., Whelan, C.T. and Williams, J. (1994) *Poverty and Time: Perspectives on the Dynamics of Poverty*, Economic and Social Research Institute, Dublin.

Pacione, M. (1982) "Evaluating the Quality of the Residential Environment in a Deprived Council Estate", *Geoforum*, 13(1), 45-55.

Page, D. (1996) *Building for Communities — The Key Factors in Ensuring Long Term Viability*, Paper presented to National Housing Conference (Royal Institute of Architects of Ireland/Department of the Environment), Waterford, April, 1996.

PAUL (1994) *The Implementation of the Third EU Poverty Programme by the PAUL Partnership Limerick*, PAUL Partnership Ltd., Limerick.

Robertson, I.M.L. (1984) "Single Parent Lifestyle and Peripheral Estate Residence", *Town Planning Review*, 55(2), 197-213.

Relph, E. (1987) *The Modern Urban Landscape*, Croom Helm Ltd, Beckenham.

Smith, D.M. (1977) *Human Geography, a Welfare Approach.* Edward Arnold, London.

11

SPATIAL PLANNING AND POVERTY IN NORTH CLONDALKIN

Brendan Bartley
Department of Geography, National University of Ireland, Maynooth

INTRODUCTION

The creation and reinforcement of community identity has been a traditional aspiration of modern town planning in Ireland. This objective underpins the *neighbourhood* principle upon which many "new towns", including Clondalkin, have been constructed since the establishment of the Irish planning system in the mid-1960s. However, it has become increasingly evident in recent times that the neighbourhood-based new towns settlement strategy has not secured social cohesion. On the contrary, it would appear that this form of spatial planning has actually served to fragment communities and reinforce social segregation, as well as making the *socially excluded* less visible, through its policies of relocating populations at the edge of the built up area.

This chapter shows how the neighbourhood unit has been used as the basic building block of the "planned" new town of Ronanstown. The spatial layout, urban design and housing management features of the "disadvantaged" North Clondalkin area of Ronanstown are examined to illustrate: (a) the extent and variation of segregation and social problems within this part of the new town, and (b) the manner in which North Clondalkin is isolated from, and its problems rendered invisible to, residents in the more prosperous parts of the city. Finally, the present situation and future prospects of North Clondalkin are reviewed in the context of the current debate about the links between social polarisation and macro-economic policy in a time of rapidly intensifying global competition.

MODERN PLANNING

Ronanstown (Lucan/Clondalkin) is one of three new "planned" towns in west Dublin built to accommodate the future growth of the city. The "new towns" were designated on the basis of a deliberate planning strategy adopted by the Dublin local authorities for the Dublin Region in the 1960s and 1970s. The new emphasis on planning and accompanying settlement strategy was linked to a change in macro-economic policies at the time. Following the introduction of the new open trade economic policies of the Lemass era in the early 1960s, the Irish Government accepted expert advice from the World Bank and United Nations about the need to introduce a new physical planning system to facilitate and regulate the changes which were expected to emanate from the newly embarked upon course of economic development (Bannon, 1989).

New planning legislation was introduced in 1963 and local authorities throughout the country were assigned the responsibility of devising and implementing spatial development plans for their areas of jurisdiction. Scope existed at the time for executive agencies other than local authorities to be given the role of overseeing the preparation and implementation of development plans by using, for example, semi-state bodies. However, this did not happen and state land use planning in Ireland thus became a purely local government function. To reflect this, local authorities were also given the formal title of *planning authorities*. To assist the recently established planning authorities with their new tasks, prominent British planners were commissioned by the Irish Government to prepare regional scale strategy plans as frameworks for the preparation of the statutory local plans. In 1964 the Government's Programme for Economic Expansion defined nine planning regions for the country. The planning consultants Nathaniel Litchfield and Myles Wright were asked to produce advisory plans for Dublin City and the Dublin Region, respectively, which would spatially articulate the national economic policies and provide the physical basis for their implementation by the Dublin planning authorities. The eminent planner, Colin Buchanan, was given the broader remit of providing a regional planning framework for the rest of the country (Bannon, 1989; Davis and Prendergast, 1995).

New Towns

For Myles Wright, the main problem facing the Dublin Region was rapid population growth. He forecast that the population of the region would increase by about 300,000 between 1961 and 1985 with much of this growth taking place as suburban development in the metropolitan area. Wright did not envisage any serious potential for increasing the residential function of the central city area and, accordingly, he prepared a plan to accommodate the anticipated population overspill from the city. He identified the major geographical constraints to future growth and, by process of elimination, suggested that the main location for future settlement growth should be to the west of the city. To the east lay the natural barrier of the Irish Sea while development to the south of Dublin was blocked by the Wicklow Mountains. Dublin Airport and its flight zone hinterland was an obstacle to northerly development. West Dublin was thus selected by default as the most appropriate area to cater for future growth (Wright, 1967).

Wright recommended the creation of four new towns in this part of Dublin linked to the four existing villages of Blanchardstown, Tallaght, Lucan and Clondalkin. The Dublin planning authorities incorporated the Wright proposals in modified form in their Development Plans from 1972 onwards. The Lucan and Clondalkin axes of Wright's scheme were amalgamated under the proposed name of Ronanstown. In summary, instead of providing four linear shaped towns to accommodate the expected 300,000 population growth, three concentric new towns were designated with target populations of 100,000 each. The village areas around which the new towns were to be built would eventually form an outer arc from the south west to the north east of the city, transforming Dublin from a small, compact high density city into a large, sprawling decentralised metropolis around a declining inner-city (Conlon, 1988). In Britain, special Development Corporations were usually established for the purposes of promoting and securing the implementation of the new town objectives. This was not to be the case in Dublin where the development of the new towns was generally allowed to occur on a *laissez faire* market basis with guidance for prospective developers about the locations of commercial and community infrastructure being provided by the Development Plans of the two Dublin planning authorities. It was envisaged that mutual and complementary planning and housing objectives could be pursued through co-operation as the two authorities shared the same chief planning officer and a joint City and County Manager (Bannon, 1989; Davis and Prendergast, 1995).

The Neighbourhood Unit
While the new towns were to be located in the Dublin County area, the City planning authority was the prime mover in the programme. They had the greatest housing needs and purchased large tracts of lands in each of the new town areas in the County. The *neighbourhood unit* was adopted as the overall design principle for the layout new towns. The neighbourhood unit had first appeared as a planning concept in the work of Clarence Perry in the United States in the 1920s. Writing about "Housing for the Machine Age", Perry proposed that housing layouts should be designed on a cellular basis which would (a) confine local vehicular access to *terminating traffic* which had genuine calls to make in the area, and (b) divert all other *through traffic* to bypass the area along major traffic roads which would form unit boundaries. Within these roads the location of shops, schools, churches and other community facilities would be decided in advance with a view to maximising access and safety for pedestrians generally and young people in particular. The underlying philosophy of this approach was that small scale arrangements of dwellings could permanently enrich community life by stimulating frequent interaction between residents and engendering a spirit of neighbourliness through spontaneous co-operation (Hall, 1989).

Perry's ideas were imported to Dublin via British planning. Very few of the planned Council housing estates built in Britain prior to 1945 incorporated Perry's ideals. Indeed, the roads layout of most pre-war housing estates tended to reflect the geometrical drawing-board patterns favoured by engineers at the time. The neighbourhood unit was endorsed by the Dudley Report on Housing Standards and the Reith Commission Report on New Towns, both of which were adopted by the British Government at the end of the Second World War. The Reith Commission proposed that residential areas should be planned in the form of neighbourhood units of 5,000 to 12,000 population, each with its own amenities (shops, community centre, primary school and chapel). The aim was that every primary school child should have a walk to school of no more than five minutes and that all housewives would have similar access to local neighbourhood shops. It was also envisaged that the new towns produced on the neighbourhood principle would help to reduce class segregation and produce socially balanced communities (Cullingworth and Nadin, 1994; Hall, 1989; Greed, 1993).

Social Engineering: Designing the Transformed City

The new towns built in Britain over the next twenty years became laboratories for testing many of the social engineering principles devised by the planning pioneers including those of the neighbourhood unit. In the period immediately after the war, the neighbourhood concept with its associated low density, low-rise housing was generally accepted as the basic building block of the new towns by planners. However, gradual experience of the growing new towns together with new data from social research and the accelerating growth of car traffic, combined to bring about a reappraisal of the neighbourhood principle. It was found to be flawed following its application in the early (Mark 1) new towns because it failed to provide the quality of urban life expected by the new town inhabitants who suffered from "suburban blues" (Prestwich and Taylor, 1990; Greed, 1993). A view developed that future (Mark 2) new towns to be successful needed vigorous, thriving and viable Town Centres rather than low density suburban style expansion. Some critics argued that planners should promote the renewal and regeneration of existing city neighbourhoods instead of dispersing urban population (Jacobs, 1965).

The anticipated *balanced communities* had also failed to materialise in the early new towns. It had been hoped that the presence of a broad spectrum of the social classes in each neighbourhood would attract a similarly balanced distribution of commercial, industrial and service employment to the area, and that the middle classes would provide potential leadership sources for the community. In practice, however, the new schemes appeared to intensify rather than reduce status distinctions. The managerial classes chose to locate away from the new neighbourhoods and opted instead for nearby "unspoilt" villages and high amenity commuter districts while single class ("rough") estates developed in many new town areas.

The neighbourhood unit was abandoned as a basic building block of the new town until it reappeared in some of the 1960s (Mark 3) new towns which were designed to facilitate the increased choice available to residents through the liberating benefits of the enhanced mobility afforded by car ownership. It was one of these Mark 3 new towns, Milton Keynes, which Myles Wright attempted to emulate in his plan for Dublin. Milton Keynes was constructed around a lattice of roads designed to enhance the accessibility of the dispersed inhabitants to the spread-out services and activities of the town. Basically, it was assumed that almost all households would have access to a car and that residents could now travel longer distances by road to avail of a wider selection of geographically dispersed services. A single, high density, multifunctional urban centre was no longer essential in the new motorisation age where freedom of choice was afforded by mass car ownership. The road network based Milton Keynes framework was also the model adopted by the 1971 Dublin Transportation Study (DTS) which was effectively a *transportation plan* follow-up to Wright's *settlement plan*.

Although the Myles Wright plan was never *formally* adopted by the Irish Government and the 1967 Draft Dublin County Development Plan took little account of his proposals, the situation changed dramatically with the 1972 County Plan. The Myles Wright strategy was now embraced for its potential to take pressure off Dublin City's growing population and reduce pressure on its radial traffic routes. It was the subsequent Plan of 1983 which made the clearest statement of the commitment to the neighbourhood concept as an integrated element of new town strategy for Dublin. Policy 2.4 of the Plan contains the following statement:

> In the case of residential development for development areas, it is Council policy to implement a strategy of neighbourhood community development based on the residential community

requirements of a population of approximately 5,000 persons. Aside from the provision of dwellings, this concept includes the provision of a primary school, church, local neighbourhood shopping and commercial facilities, sites for a community centre and youth club and a local park. It is policy to group these facilities in order to provide a focal point for each neighbourhood (Dublin County Development Plan, 1983, pp. 53-54).

In short, the 1983 Plan attempted to produce a Dublin version of Milton Keynes by marrying Perry's neighbourhood idea to Wright's new town overspill philosophy and the associated road network proposals contained of the 1971 DTS. The separate neighbourhood communities would be integrated by the provision of additional services (based on higher levels of population) at central locations which would incorporate the catchment population of the aggregated neighbourhoods. In the case of the three new towns, the highest population aggregate would obviously be the new town itself and the highest level services and facilities available in each would be located in a *town centre* at the heart of the new town.

Assessment of the Dublin New Towns Strategy
In 1996, none of the three new towns has achieved their 1991 target population of 100,000 (Figure 11.1) and Tallaght is the only one with an operating town centre. A town centre complex is currently under construction in Blanchardstown. Ronanstown (Lucan/Clondalkin) has the second highest population of the three new towns but it appears that it will not now have a centrally located Town Centre due to the recent rezoning of, and granting of planning permission for, an alternative edge of town site to allow a major regional shopping and commercial complex to be developed at a geographically strategic node adjacent to the junction of the upgraded Galway Road (N4) and the Dublin Orbital Motorway (M50).

It may be too early to judge whether or not the Dublin new towns have been a success. Critics can certainly point to the time lag involved in providing them with Town Centres and other essential services. Perhaps these facilities would have been available earlier if special, executive Development Corporations similar to those in the UK had been given the task of ensuring their success instead of leaving their development to the vagaries of the private market. However, this is a criticism of the approach adopted to implementation rather than of the need for, or objective of providing, new towns *per se*. Perhaps the new towns were also built too close to the Metropolis to become autonomous, self-contained entities in the short term (Dublin Transportation Task Force, 1986).

Figure 11.1: Population Growth, 1961–89

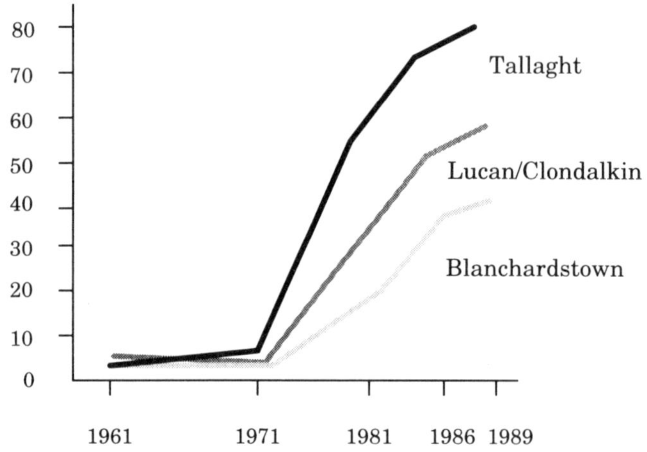

[Line chart showing population growth from 1961 to 1989 for Tallaght (reaching ~80), Lucan/Clondalkin (reaching ~58), and Blanchardstown (reaching ~42). X-axis: 1961, 1971, 1981, 1986, 1989. Y-axis: 0 to 80.]

Source: Dublin County Council Planning Department

At a superficial level, the aim of achieving social balance in the new towns appears to have been achieved. Figure 11.2 suggests that the social class profile of the new towns compares favourably with distributions for other single class areas in Dublin. However, this global picture of the social composition of the new towns is seriously misleading insofar as it masks significant internal patterns of socio-spatial segregation. Far from creating balanced communities the new towns may have accentuated social segregation by contributing to the creation of new ghettos — areas of physical marginalisation as well as social and economic exclusion — in the suburbs. This issue will be explored further here through an examination of the case study of North Clondalkin in Ronanstown but many of the points made could apply with equal validity to problem areas in the other new towns.

town but many of the points made could apply with equal validity to problem areas in the other new towns.

Figure 11.2a: Age Structure — Dublin

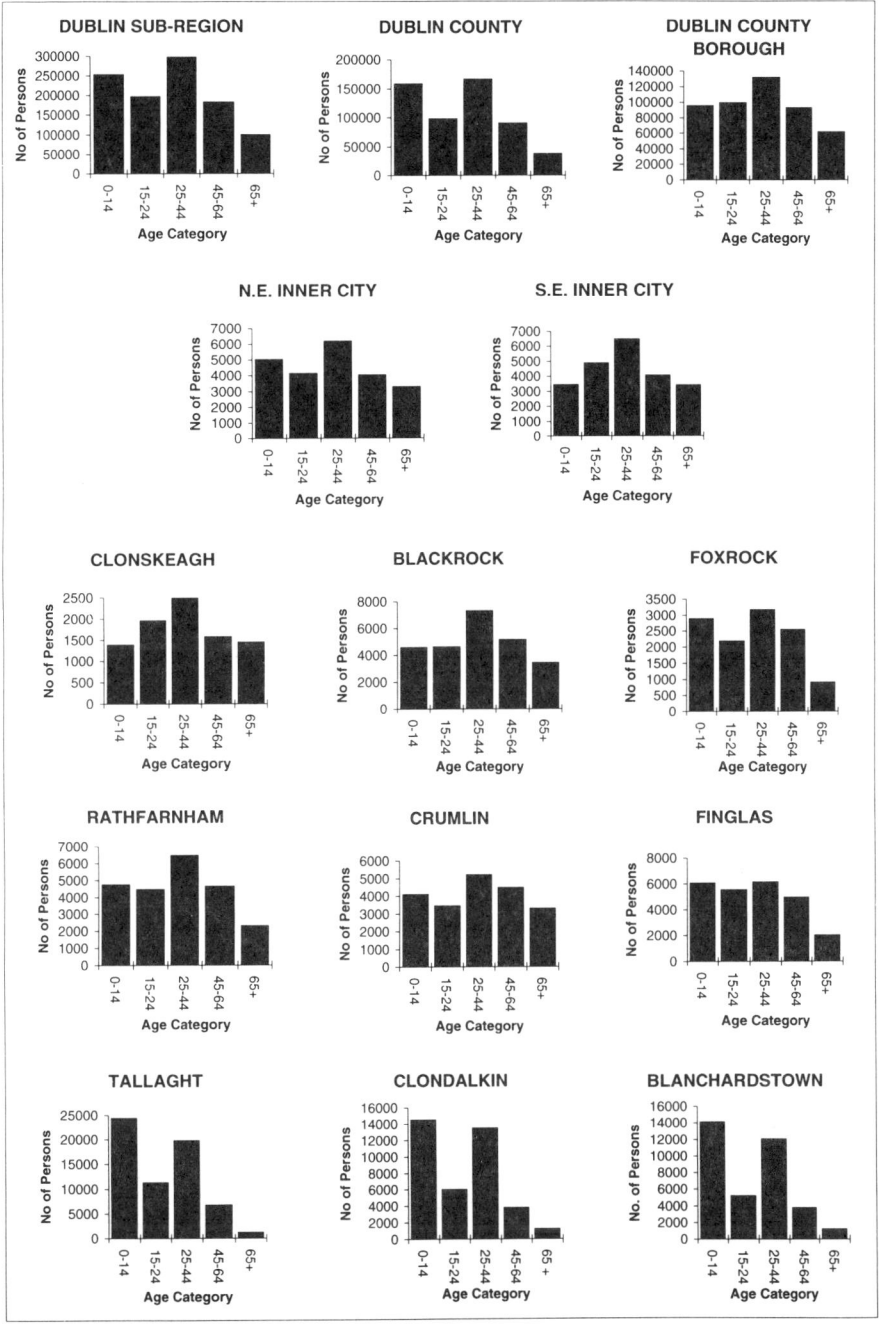

Copyright - B. Bartley and C. McHugh (1997)　　　　**Source : Census of Population 1991**

Figure 11.2b: Social Class Structure — Dublin

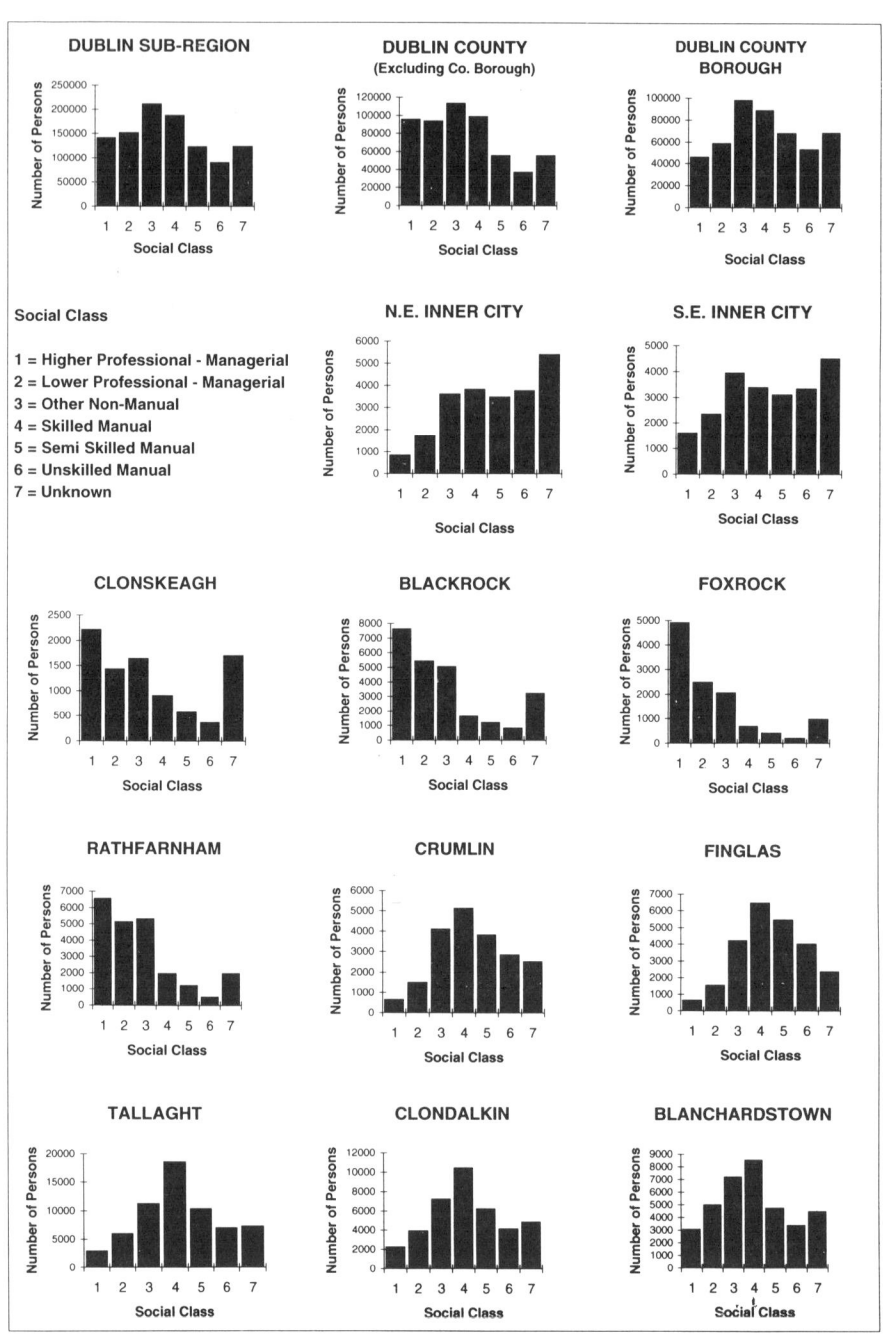

Copyright - B. Bartley and C. McHugh (1997)

Source : Census of Population 1991

CASE STUDY: NORTH CLONDALKIN

North Clondalkin contains the three neighbourhoods of Neilstown, Rowlagh and Quarryvale (see Figure 11.3). Its boundaries are formed by the Western Parkway Motorway (M50) to the east, the Lucan By-Pass (N4 Maynooth/Galway Road) to the north, the Balgaddy/Fonthill Road to the west and the Grand Canal and Cork railway line to the south. There are approximately 3,600 houses in the area with very little variety in the type or size of housing available: virtually all of the houses in the area are three bedroom, two storey buildings.

What one sees if one ventures into North Clondalkin are littered and unkempt approach roads, run-down neighbourhood centres, public buildings which are invariably surrounded by large palisade fencing and shuttering, poorly kept open spaces, a skyline dominated by large electricity pylons and housing estates which face inward and turn their back on the public areas. The neighbourhood centres which were to be at the heart of each neighbourhood have failed. They are run-down and often contain many empty units. In the meantime, the villages of Lucan and Clondalkin, located at the edges of the proposed new town, have become thriving shopping and business centres while North Clondalkin, which was intended to be at the heart of the new town, is effectively isolated from both villages and is almost devoid of facilities (Clondalkin Partnership, 1996).

The population of the area is estimated by the 1993 North Clondalkin Task Force Report to be about 16,300. This figure includes an estimated 450 travellers who live in unofficial halting sites in the area. Table 11.1 presents the comparative age structure of the area relative to both Dublin City and County and the state. It confirms that North Clondalkin has an unbalanced age structure comprised essentially of young to middle aged parents with young children. More than 30 per cent of the population are in the 25-44 age group and over 40 per cent are less than 14 years of age.

> The age structure is just one reason why North Clondalkin poses particular social planning problems. As a new community with little history it has not inherited any ready made solutions or structures to cope with the demands confronting it now and in the future (Collins and Crowley, 1993).

Figure 11.3: Action Draft Plan, Neillstown, Rowlagh and Quarryvale

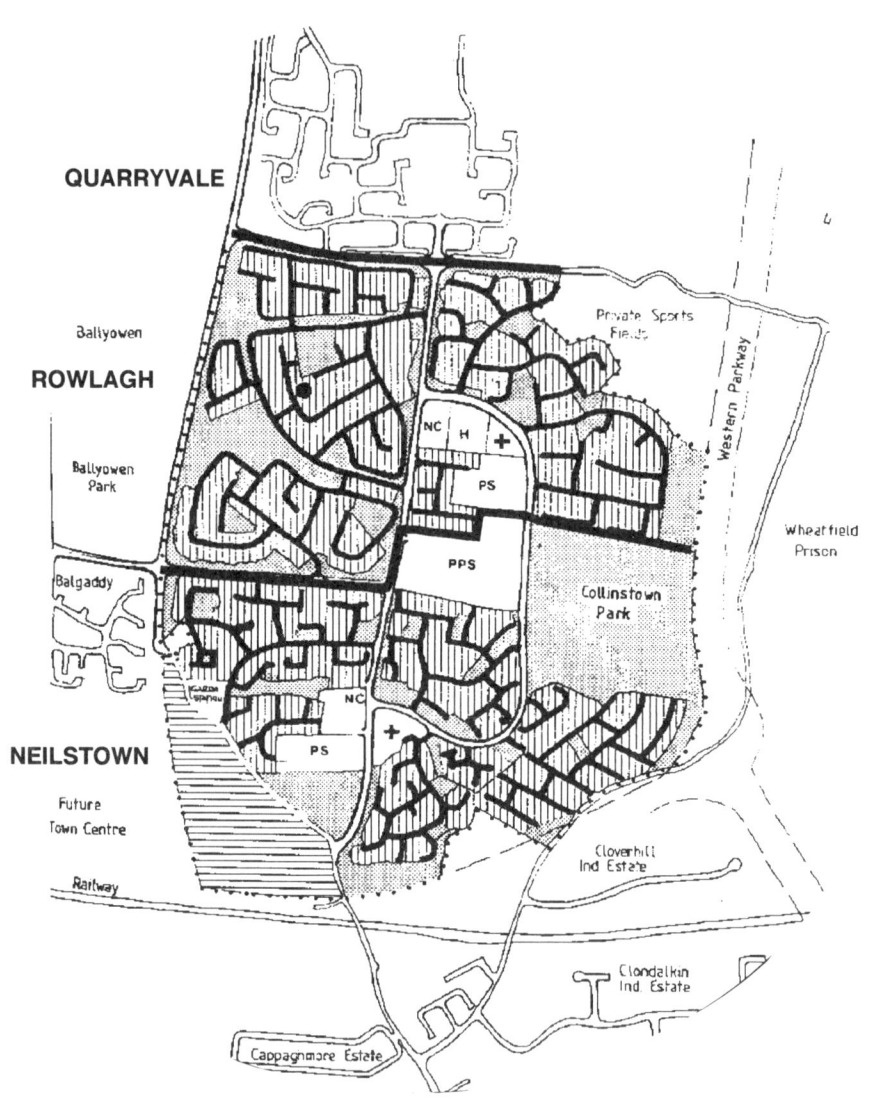

Table 11.1: Age Structure — Ireland, Dublin North, Clondalkin

Age Bands	Republic of Ireland (%)	Dublin City & County (%)	North Clondalkin (%)
0-14	27	25	42
15-19	9	10	10
20-24	8	9	6
25-44	27	29	32
45-54	10	10	6
55-59	4	4	1
60-64	4	4	1
65 +	11	9	2

Source: Census of Population, 1991

Economic and Social Marginalisation: Some Indicators of Exclusion

North Clondalkin is in many respects on the wrong side of the tracks. The railway line is a major boundary severing it from Clondalkin Village and from the extensive private housing areas of both South and East Clondalkin as well as the County Council housing estates which predominate in the neighbourhoods of West Clondalkin. The North Clondalkin neighbourhoods consist mainly of public housing estates built by Dublin Corporation since the mid 1970s alongside a smaller number of private estates (See Table 11.2). Precise information about the socio-economic structure of the area based on occupation is difficult to extract from the census because the enumeration districts used for North Clondalkin include parts of West Palmerstown and South Clondalkin, both of which are known to contain a relatively high proportion of residents who fall into the professional and managerial classes. The high percentage of local authority housing is probably the best general indicator of the social class structure in North Clondalkin. Local authority houses account for 75 per cent of all housing in the area.

Table 11.2: North Clondalkin Neighbourhoods and Estates

Neighbourhood/ Estate	Year Built	Owner-ship	PercentageTenant Purchase/Sales Schemes
Neilstown			
Neilstown	1977	Corporation	49.3
Moorfield	1977	Corporation	44.0
St. Ronan's	1979	Corporation	34.3
Palmerstown Woods	1979/80	Private	n/a
Wood Avens	1984/86	Private	n/a
Collinstown	1986	Private	n/a
Total			41.6
Rowlagh			
Rowlagh	1977	Corporation	48.4
St. Marks	1979/85	Corporation	25.1
Harelawn	1982	Corporation	19.5
Liscarne	1991	Corporation	14.8
Wheatfield	1979/80	Private	n/a
Glenfield	1979/80	Private	n/a
Oatfield	1980/84	Private	n/a
Total			30.2
Quarryvale			
Greenfort	1984	Corporation	29.7
Shancastle	1985	Corporation	18.3
Total			24.7

Note: n/a = Not Applicable

Neilstown, Rowlagh and Quarryvale are all included in South Dublin County Council's 1994 "Areas of Need" (CODAN) Study. The CODAN Report provides information for a wide range of indicators of disadvantage, including details about income and unemployment rates for principal earners. Table 11.3 summarises its findings in respect of some key indicators for the North Clondalkin Neighbourhoods. The socio-economic profile of the area depicted by the CODAN Report indicates a population experiencing generally low levels of income and high levels of unemployment and dependency. It is particularly noteworthy that over 80 per cent of principal earners in the area fall below the national mean average weekly disposable income of £198. The report confirms what was already known, that the neighbourhoods of North Clondalkin fall way below the national norms in terms of income, status and job security. The absence of a major employer either in or adjacent to the area compounds the effects of marginalisation experienced by the community (Collins and Crowley, 1993).

Table 11.3: Profile of Areas of Need

Indicators	Neilstown	Rowlagh	Quarry-vale
Population	4,523	6,855	2,516
Average Household Size	4.0	4.38	4.5
% Households with 6 or More	18.7	22.9	24.3
Primary School Enrolment	632	886	935
% Lone Parent Households	22.7	23.8	27.3
% Private Housing	41.4	35.4	0.0
% Tenant Purchase	1.6	30.2	24.7
% on Transfer List	17.9	22.3	32
% Principal Earners Unemployed	55.9	73.4	59.5
Average Gross Income*	£123.00	£121.93	£131.72
% Principal Income < £150 gross	75.4	77.6	73.4
% Principal Income > £200 gross	7.3	5.6	10.2

Source: Codan Report, Vol 11, 1994.

* Average adult industrial worker earnings 1993 = IR£290.14.

North Clondalkin occasionally features in the wider media. When this occurs, it is usually as a result of some major law-and-order "disturbances" in the area. Homicides, drugs marches and riots, and assaults on fire brigade and garda vehicles have been among the more prominent events to have reached public attention through the national print and broadcast media in recent years. Perpetual problems such as "joy-riding" receive little media coverage unless they are associated with a catastrophic outcome, such as a crash involving death or serious injury. The gardai are under instructions not to pursue stolen cars into a number of the estates in the North Clondalkin neighbourhoods.

A report on *Urban Crime and Disorder in North Clondalkin* was produced in 1992 by a Government interdepartmental group following a number of media reports on major calamities in the area. This report contains very little contextual information about North Clondalkin and no comparative or historical crime data. Not surprisingly, therefore, the analysis section of the report is weak. It does, however, supply an extensive range of conclusions and recommendations. One proposal considered by the gardai but not included in the report was the feasibility of installing a closed circuit television (CCTV) system for the purpose of monitoring potential criminal activities on the approaches to and within the area particularly along the central road spine (Neilstown Road). A number of the suggestions for improvement which appear in the report concern specific aspects of law and order, health care, education and environmental matters involving various bodies in the Public Sector. There is an implicit acknowledgement in the report that North Clondalkin is a marginalised area which has been neglected by the public authorities.

Another North Clondalkin Task Force Report, commissioned by the local communities themselves, goes further in its analysis of the difficulties facing the area. It points out that the failure to provide the planned Town Centre for Ronanstown coupled with the fall-off in housing construction in the area since the mid-1980s (only about half of the allocated space has been built upon), has seriously impinged on the quality of life in the isolated area of North Clondalkin.

> The area remains in something of a planning limbo, lacking amenities, industrial, recreational and commercial infrastructure and embodying a general sense of being unfinished. (Collins and Crowley, 1993, 7).

Transportation: Exclusion, Entrapment and Fragmentation

The physical isolation of North Clondalkin is ironic given that it is defined geographically by major road networks. The only available direct access to the villages of Lucan and Clondalkin for the residents of North Clondalkin is provided by the Fonthill-Newlands Road which crosses a narrow (single lane) bridge over the railway line to the south east of Neilstown. In reference to this geographical divide, the local community information newsletter has commented:

> It's only a 19th Century crossing over some railway tracks, yet it must rank as one of the great physical dividers of our time. Not as internationally famous as the Allenby Bridge across the River Jordan but every bit as divisive (The Buzz, November 1994).

The provision of a planned internal road which it is claimed would unite Clondalkin (the Fonthill Road extension) has until recently remained low on the Council's list of priorities for new road spending. While the construction of this road might improve the situation for pedestrians and public transport services in the area, such benefits are not self-evident as the road is designed to cater not for them but for private and commercial traffic. Car owners would certainly be amongst the chief beneficiaries of the new link road. However, it would not significantly enhance accessibility by means of private car for most residents of North Clondalkin. "Car ownership is much lower in the area than nationally, reflecting higher levels of unemployment and poverty. 67 per cent of households in Rowlagh for instance are without a car" (Clondalkin Partnership, 1996, 20).

This raises some further ironic points about the layout of the new town neighbourhoods in North Clondalkin: in addition to being defined by road boundaries, the housing estates of which they are comprised are intentionally designed to cater for the needs of the private car. The estates with most roadspace are the local authority estates. These of course have the lowest levels of car ownership. Viewed from aerial photographs, the cul-de-sac style layouts of the public housing areas display far more imagination and design variation than the straight-line rows of housing which are more typical of the private estates. The fundamental influence on the layout patterns in all cases is the roads hierarchy with roads at each level feeding sequentially into higher order networks to provide a widening range of access for motorists. Yet, because of their low car ownership rates, the households with the highest percentage of local roadspace (i.e. local authority estate residents) are least able to take advantage of the potential of the wider roads system. Moreover, the resulting reliance on public transport does not provide much compensation for this deficiency. Sprawling, low-density, low-rise, cul-de-sac style suburbs provide the worst possible operating environment for urban bus services. This is reflected in North Clondalkin where buses are not free to enter and wind their way around the various estates to collect passengers with the result that the available service is confined to just two main routes (Neilstown Road and Coldcut/Balgaddy).

The assumptions of the Milton Keynes model that residents would have the freedom to avail of services and facilities spread over a wide geographical area do not apply to the imported version implemented in North Clondalkin. Instead of an expansive outward linking orientation the area has the restricted inward focus of an isolated enclave. This *implosive* quality is reinforced by the fragmentation effect created by the internal design of the neighbourhood housing estates with their emphasis on separateness: separate neighbourhoods, separate estates, separate house clusters and cul-de-sacs. In the absence of opportunities for social interaction due to the limited service provision and the anti-pedestrian bias of the roads oriented estate layouts, this accent on diminishing, small scale identity militates against the development of social cohesion in the area. Internal differentiation and competition is stressed at the expense of local integration and area unity.

The Internal Geography of Exclusion: Housing Status and Segregation

The CODAN report presents aggregate results for neighbourhoods which mask internal differences between the neighbourhood estates. Apart from the clear-cut distinction between private and public authority estates there are very definite distinctions between the local authority estates. There is a spatial pattern and logic to the apparent differentiation. The ratio of public to private housing in the three neighbourhoods increases with distance (and isolation) from Clondalkin Village. Neilstown, the neighbourhood nearest to the village, has 42 per cent private housing. This compares with 35 per cent for Rowlagh which is located between Neilstown and the most northerly neighbourhood Quarryvale which has no private housing. Moreover, the local reputation of the public housing estates appears to be positively correlated with proximity to Clondalkin Village. The status of the local authority estates diminishes as one proceeds northwards away from the railway bridge which connects North Clondalkin with the Village.

This directional pattern is also apparent with the variation in visual quality of the estates: there is a progressive decline in the appearance of the local environment as one moves northwards, with the interiors of the local authority estates closest to the Lucan Bypass showing the greatest evidence of degradation and neglect. Interestingly, the appearance of the housing terraces fronting onto the roads which bypass these estates are usually amongst the finest in the locality. However, the visual amenity of the housing and its surrounding environment disimproves as one penetrates deeper into the estates. Once again, the best is on the outside while the worst is hidden away from public view. A survey by the author of one edge-of-estate street at Shancastle in Quarryvale revealed that all but one of thirty houses facing onto the main road were being purchased by the occupiers from the local authority. This is way above the prevailing 18 per cent average rate of tenant purchase for this 240 house estate as a whole. With one exception, the principal earner in all of the households in the surveyed street was in paid employment. Again, this is totally at variance with the general pattern of income for an estate where less than 27 per cent of principal earners are in paid employment. The local authority housing departments do not appear to have formal policies about entitlement to housing at specific locations within estates. However, the extent of the above clustering patterns along the boundaries of the estate suggests that some sort of housing selection, or allocation, mechanism is at work in this instance.

The private housing estates in North Clondalkin are located along the outside boundaries of the three neighbourhoods. It is the private estates which abut the most expansive areas of open parklands in North Clondalkin. The value of private houses, as determined by price on the second-hand housing market, diminishes with distance from Clondalkin Village. The more *northerly* the house is located in North Clondalkin, the less it is likely to fetch on the open market. The private housing estate nearest to the village and shown on all maps as part of the neighbourhood of Neilstown is, in fact, physically separated from the rest of the neighbourhood. This estate, Palmerstown Woods, is walled off from the surrounding local authority estates and has a separate, external link road to Clondalkin Village. Despite having the closest proximity to Clondalkin Village of all the estates in North Clondalkin, the claim to separate identity of this estate is asserted further through its being named after an entirely different, higher status village (Palmerstown) in the west Dublin area.

The recent resurgence of sales in the private housing market in Dublin has not given rise to further house construction in North Clondalkin. New low to medium cost residential development in West Dublin has tended to occur instead at the Lucan end of Ronanstown. Estate agents in Lucan informed the author that most new housing in the vicinity is sold on the basis that the Lucan area is distinct from, and clearly unattached to, North Clondalkin. Purchasers feel that they are buying into a more exclusive area which, despite its proximity to North Clondalkin, is clearly separated from it in terms of status; but also geographically because of the absence of connecting roads and the presence of an undeveloped buffer zone between the two areas. An attempt in 1990 to have the formal single town (Ronanstown) development strategy overturned and replaced by two autonomous town developments was unsuccessful. Nevertheless, the sense of exclusivity in Lucan with regard to North Clondalkin still prevails and is reinforced by the fact that virtually all of the new housing complexes have been provided with protective boundary fencing which usually takes the form of ornate railed walls and gateways which restrict entry to the estates.

The Wider Geography of Exclusion: Physical Isolation and Invisibility

If their social and economic circumstances together with the isolated location and internal design of their neighbourhoods combine to entrap many local residents in North Clondalkin, these same features also serve to keep outsiders away from the area. It requires a special effort to enter North Clondalkin. The structure of the roads network hierarchy is such that anybody who does not need to visit the area for a specific purpose will automatically bypass it. Of course, it is unlikely that residents from more prosperous parts of the city would seek out the limited range of facilities in North Clondalkin in preference to the presumably more extensive range available in their own localities. The "problem reputation" of North Clondalkin is another factor likely to deter potential "tourists" from visiting the area. Even when it is entered by outsiders, penetration is limited by the cellular layout structure of the neighbourhoods and the restricted access of the local authority housing estates. It is usually not possible to drive directly through these estates: motorists entering the maze of cul-de-sacs in one estate must find their way out again before they can enter the next such maze in the adjacent estate. There is, therefore, no inducement to enter a particular estate unless one has a definite reason to be there. In short, the overall effect of the location and spatial design of North Clondalkin is to make less visible the social exclusion experienced by the isolated and fragmented communities of the area and to obscure from general view the specific problems of its inhabitants.

In 1844 a German, Friedrich Engels arrived in England. He travelled to Manchester and immediately made an effort to understand its overall structure — how each part related to the other. Making connections, seeing both the details and the whole, Engels wrote one of the first accounts of what would later be called the "sociological view" of the city. He noted that the wealthy were seldom obliged to see the real living conditions of the poor labouring districts:

> The town itself is peculiarly built so that someone can live in it for years, travel into it and out of it daily, without ever coming into contact with the working class quarters or even with the workers. I have never come across so systematic a seclusion of the working class from the main streets as in Manchester. I have never elsewhere seen a concealment of

such fine sensitivity of everything that might offend the eyes and ears of the middle classes (Engels, 1969, 80).

Engels ventured into the back streets to see for himself the living conditions of the poor. He was shocked at what he saw and expressed his amazement that "such a district exists in the very centre of the second city of England, the most important factory town in the world" (Engels, 1969, 86). The description provided here of North Clondalkin has many similarities with the inner city ghetto described by Engels. It is easier today to be aware of the problems of inner city areas because of their immediate proximity to busy commercial and employment districts. The problems in the suburban areas are harder to see. Areas like North Clondalkin appear to be the relocated equivalent of old industrial inner city problem areas; now moved to the "invisible" suburbs of the new towns away from, and out of the sight of, residents and workers in the more prosperous parts of the city.

In spite of the isolating effects described above, and despite evidence in the CODAN survey results that many respondents in the disadvantaged neighbourhoods wish to move away from the housing estates in which they now live, local residents are reported by the planners to be satisfied with the neighbourhood approach to the development of the new towns. A planning survey carried out by Dublin County Council in 1986 as part of the Development Plan Review sought to establish the extent to which the application of the neighbourhood concept by the Council had been successful in physical and community terms. The survey consisted of interviews with residents of eight neighbourhoods, two of which (Neilstown and Rowlagh) are located in North Clondalkin. The general findings of the report were that the neighbourhood approach to the development of the new towns had worked satisfactorily. Overall, it was found that, "taking everything into account", most residents were satisfied with their neighbourhood. In particular, residents expressed a high degree of satisfaction with the *design layout* (emphasis added) of their local areas and neighbourhoods (McCarron, 1988, 4). This paradoxical situation can be explained by a critical review of the role of planning in the shaping and reshaping of modern cities.

PLANNING AND SOCIAL ENGINEERING

In broad terms, urban planning is thought to be concerned with the co-ordination of land-uses and development activities. It professes to ensure that future development is rational and that it does not unnecessarily damage or waste architectural, environmental or community resources. Beyond this, it is difficult to pin down what exactly planning is about. The "role" of planning and the range of purposes it ought to serve are matters of continual debate. Within this debate, however, the justification for the intervention of planning is usually attributed to the concept of the "public interest". Planners are typically portrayed as politically neutral, technical experts who serve the public by providing advice about the best use of state resources and warning against potentially damaging developments. However, like most human activities, planning can have positive and negative outcomes in that different groups experience its results as beneficial or detrimental to their interests.

There is a growing literature illustrating the way in which the social engineering side of planning has, through the introduction and use of land zoning and development controls, "inadvertently" contributed to the problems of residential segregation by allowing the wealthy to distance themselves from the problems of the poor (see Hartshorn, 1992; Gerckens, 1994; Knox, 1994). Urban planners are also often depicted in the literature as urban managers or social engineers (Pahl, 1977; Kirk, 1980; Saunders 1986; Knox, 1995) and as professing to be concerned with issues of social justice (Greed, 1993; McGuirk, 1996). As such, one would expect them to be concerned about the issues of marginalisation and social exclusion which affect places like North Clondalkin.

Urban Problems and Urban Analysis

Table 11.4 summarises the various approaches which have been developed to explain and tackle the social problems associated with marginal urban areas. Most of these approaches assume that the causes of the problems are located within the areas themselves or the individuals and groups who occupy them. Others attribute the problems to institutional malfunctioning or political manipulation and assert that they can be remedied by appropriate

Table 11.4: Differing Explanations of Urban Problems (after Bennington, 1979)

Theoretical Model of Problem	Exploration of the Problem	Location of the Problem	Key Concept	Type of Change Aimed For	Method of Change
Culture of poverty	Problems arising from the internal pathology of deviant	In the internal dynamics of deviant groups	Poverty	Better adjusted and less deviant people	Social education and social work treatment of groups
Cycle of deprivation	Problems arising from individual psychological handicaps and inadequacies transmitted from one generation to the next	In the relationships between individuals, families and groups	Deprivation	More integrated self-supporting families	Compensatory social work, support and self-help
Institutional malfunctioning	Problems arising from failure of planning, management or administration	In the relationship between the "disadvantaged" and the bureaucracy	Disadvantage	More total and coordinated approaches by the bureaucracy	Rational social planning
Maldistribution of resources and opportunities	Problems arising from an inequitable distribution of resources	In the relationship between the underprivileged and the formal political machine	Underprivilege	Reallocation of resources	Positive discrimination policies
Structural class conflict	Problems arising from the divisions necessary to maintain an economic system based on private profit	In the relationship between the working class and the political and economic structure	Inequality	Redistribution of power and control	Changes in political consciousness and organisation
Social polarisation	Problems arising from marginalisation of "weak" groups	In the relationship between "marginalised" groups and political and economic structure	Exclusion	Inclusion of marginalised groups in political, social and economic decision-making processes	Creation of for and structures to secure involvement of marginalised in decision-making processes

corrective action. I believe that the role of planners as urban managers and social engineers can usefully be understood in terms of the wider social system emphasis which is provided by the structural conflict approach to urban analysis; and that is the approach which is adopted in this chapter.

For structuralists it is not so much planning decisions as the underlying pressures associated with business competition and profits which determine the shape of towns and cities in modern market economies. In this view, planning plays an important, but secondary, supportive role in the shaping of the built environment. The emerging spatial arrangements, or city shapes, which facilitate competitiveness and profitability are those which planning, in the "public interest", is expected to secure.

The patterns produced are, of course, complex and dynamic because in an open market system of intensifying, globalising economic competition, the unrelenting application of more efficient technology and innovative practices creates a continuous need for appropriate new urban environments. Thus, cities in the twentieth century have a very different appearance to their predecessors. They are characterised by the development of high rise offices in the Central Business District, decentralisation of industry and residential suburbanisation — all of which can be shown to be important facets of the modern market economy (Fainstein, 1994; Knox, 1994; 1995). The fluctuating urban patterns facilitated at different periods of city growth in this century have included urban intensification (high density, high rise development), de-urbanisation (decentralisation/suburbanisation) and re-urbanisation (inner city renewal). All of these changes have been accompanied by increased residential segregation (Bartley, 1995).

Town Planning: Managing Transformation and Problems

Planning as a state activity has played a role in smoothing the way for the emergence of these new city shapes or patterns. It manages these transformations (through Development Plan zoning and control policies) so that future change is "rationalised" and undesirable change restricted where necessary. It also manages the potential conflicts which arise in the course of these transformations. In particular, urban planning plays a particularly important role in fragmenting, or *atomising*, the labour force while simultaneously defusing conflict by ameliorating their living conditions. Urban growth and change helps to promote profitability and competition for the business sectors. However, by spatially concentrating the urban populations as labour and consumer markets, urbanisation also increases the potential for a united opposition to the prevailing social and economic arrangements. This prospect is pre-empted by the state through various forms of control including the planning and management of segregated housing markets and sub-markets for people with low incomes (Bartley, 1995).

Thus, the widening social and geographical distance associated with the growing polarisation of the wealthy and poor is one trend that accompanies city expansion in modern market economy societies; another is the increasing residential segregation attributable to the fragmentation of the labour force (including the unemployed) in an increasingly sub-divided housing market. Housing is a commodity which can be produced and exchanged for profit and, as such, the private housing market is part of the functioning market economy. However, housing also features as a welfare function supplied to the poor by the state, which has the dual role of (a) regulating conflicts between competing business sectors, and (b) mollifying the labour force by ensuring that it receives basic essential services. State housing, therefore, plays an essential role in pacifying and stabilising the existing workforce. The creation of an urban environment based on homogenised but distinctive and competing residential communities inhibits the labouring class occupants from recognising their common interest in the economic system. They are encouraged to identify with their respective communities and to compete with each other on a local scale which undermines their potential to develop a sense of solidarity and their ability to seriously challenge the existing rewards system. This, together with the activities of other state controlled institutions (e.g. education, the legal system, the media), serves to nullify latent opposition to the existing economic arrangements thereby reinforcing the authority of the dominant groups who benefit from it.

Planning is implicated in this process but in the guise of a neutral technical arbiter. Its biased influence on the distribution of resources is disguised behind its supposedly apolitical mask of serving the "public interest". Applying these arguments to North Clondalkin, we can observe that one of the main tasks of modern planning has been to accommodate expanding business in the city centre together with the resolution of prospective housing crises in the city. This has involved the breaking up and dispersal of established communities and the creation of new dispersed and weakened communities based on the concept of the *neighbourhood unit*. This idea has the objective of "recreating" community identity by dividing urban areas up into the smallest possible groups consistent with the efficient delivery of state provided essential services such as roads and schools.

The neighbourhood, as we have seen in the case of North Clondalkin, is usually constructed around the provision of a primary school and local shops which are within a few minutes walk of all households in the unit and will usually have a population which falls within the required "efficiency" threshold of 5,000 to 12,000 people. A secondary school and other higher order services can be provided *efficiently* in an "amalgamated" area consisting of two to four such neighbourhoods. The dispersed and segmented population catchments of these amalgamated areas can then be "combined" to justify the provision of further higher order facilities and so on. Thus, "improved" living standards are provided, in the most cost-effective way, for the relocated population who are both pacified and fragmented in the process. The reality in many places, such as North Clondalkin, is not an integrated larger community but an isolated labyrinth of differentiated housing enclaves. *Integration* in such cases only applies to the population figures used to calculate the optimal location of necessary services.

In summary, the creation and reinforcement of community identity has been a traditional aspiration of modern town planning in Ireland. This objective underpins the *neighbourhood* principle upon which many "new towns", including Clondalkin, have been constructed since the establishment of the Irish planning system in the mid 1960s. However, it has become increasingly evident in recent times that the neighbourhood-based new towns settlement strategy has not secured social cohesion. On the contrary, it would appear that this form of spatial planning has actually served to fragment communities and reinforce social segregation, as well as making the *socially excluded* less visible, through its policies of relocating populations at the edge of the built up area.

Local Problems, the State and Wider Economic Forces

The structuralist approach employed in this examination of the planning and development of North Clondalkin throws an extremely negative light on the problems facing the marginalised communities in the area. Urban planning, in this view, was involved in the creation of North Clondalkin and is associated with many of its *secondary* problems of isolation, fragmentation and invisibility. Local government planning is also unable to solve the *primary* problems of unemployment, deprivation and exclusion created by the macro-economic policies decided at central government level. Such problems are a common consequence of macro-economic policies which promote and support competitiveness and profit maximisation in open market economy conditions. The processes of economic restructuring underway since the early 1970s have been accompanied by an homogenisation of macro-economic policies across the nation states of the developed world. Increasingly, as the integration of foreign exchange markets and the globalisation of finance capital links together their economies and leaves them exposed to the potential "flight of capital", many national governments have replaced *Keynsian* job-creation, growth-oriented strategies with *monetarist* macro-economic policies designed to control inflation at the expense of their job creation and economic growth priorities (Leyshon, 1995). Paradoxically, in an era of unprecedented work restructuring, unemployment and job insecurity this has contributed to a re-defining of the role of the state and the emergence of powerful neo-liberal political ideologies which advocate that government can best help to provide new jobs by letting business competition take its course. Central government is urged to be more supportive of the business sector and to further assist by: (a) relinquishing control over public resources (privatisation) ; and (b) adopting a non-interventionist approach to market competition (deregulation). Where the central state adopts this ideology, the locally experienced effects of (and obligations associated with) the consequent economic, social and environmental restructuring is typically the responsibility of local government.

Ireland is a highly centralised state in which local government has virtually no political power or autonomy. Town planning is part of the Irish local (*not* central) government system. Area-based planning at this level cannot provide definitive solutions to the problems created by the wider socio-economic system when it is itself a subordinate part of that system. It is especially ineffective at addressing local problems where it is obliged to protect and advance the interests of the wider system. Urban planners in Irish local authorities are in an invidious position. Criticisms of the localised outcomes of macro-economic policies are typically directed not at the instigators and upholders of these central state policies, but at the local authority planners who, as we have seen, play a mainly supportive and ameliorative role in implementing them. The planners, therefore, are *fall-guys* or *piggies-in-the-middle*, providing a safe buffer zone for those business interests and central state brokers who actually influence and make the key economic decisions which ultimately affect local areas.

CONCLUSION

As the divergence between rich and poor grows in the western economies, we now have a growing debate about the development of an urban "underclass" for whom the drugs trade and crime increasingly offer the best opportunities for advancement (Musterd, 1994; Mingione, 1996). The circumstances of this "underclass", and the deprived neighbourhoods in which they live, did not just invent themselves. They are a by-product of the economic restructuring engendered by free market forces and the social and economic policies to which national governments subscribe (Barrf, 1995; Hutton, 1996). The pressure on national governments in an increasingly open, and therefore competitive, international arena is to implement market regimes and labour controls that will attract mobile investment and promote international trade in goods and services. The resultant emphasis on business *shareholder* interests at the expense of community *stakeholder* concerns, produces the social polarisation and segregation which translate at their extremities into the secured fortresses (gated communities) of the *included* wealthy and the marginalised reservations (hidden ghettos) of the *excluded* poor (Davis, 1992; Painter, 1995;

Hutton, 1996). Meanwhile, the wealthy increasingly choose to provide their own private pension and insurance cover and opt out of public welfare schemes so that the welfare state and the wider community consensus it embodies is continuously eroded. The diminishing contributions of the wealthy also undermines the ability of the state to provide an adequate welfare safety net for the unemployed as well as the chronically insecure workers and the new poor (low waged, contracted and casual labour) created by the economic restructuring. A negative cycle is engendered whereby those who have well paid employment increasingly complain about the "welfare" drain on the exchequer (and their tax contributions) arising from the growing demands associated with the increased poverty. The welfare state becomes a target of criticism. An increasing reluctance to subsidise or support the besieged welfare regime translates into discussions which define welfare, rather than poverty and unemployment, as the problem to be addressed. Where the debate involves calls for the replacement of welfare by workfare — with its emphasis on personal responsibility and the need for work incentives — unemployment and poverty are redefined as individual, personal problems (relating to some lack of *will* to work) rather than structural ones (to do with lack of, and poorly paid, jobs) (Painter, 1995; Peck, 1996).

Structuralist analyses typically assume that the actions of individuals and groups are not autonomous, but are shaped by the imperatives of an economic system which is driven by competition, profit accumulation and technological change. It has been criticised for its failure to make allowances for the scope and ability of small, independent groups to influence urban change. Many of the newer area-based approaches to local development are more flexible in this regard. They fall into the category of approaches designed to address the problems of *social exclusion* (see Table 11.4). Conceptually, the social exclusion focused strategies claim to combine the strengths and avoid the weaknesses of the "failed" older approaches to tackling poverty (Room, 1993). At a practical level, they enhance the prospects of implementing grass roots, bottom-up initiatives through the involvement of local groups in partnerships that link them with private enterprise, local government and state sector agencies (Nexus, 1996). There has been much experimentation with local development initiatives over recent years in Ireland which has been favourably reviewed by the Organisation for Economic Co-operation and Development (OECD). Fintan O'Toole, in a recent review of an OECD evaluation by Charles Sabel (1996) of area-based partnerships, attributes the success to these partnerships to the fortuitous coincidence of their flexible, middle-way approach ("between free market anarchy and statist inertia") with the emergence of new decentralised and flexible business models in the global economy. However, as Walsh (1999) has pointed out elsewhere virtually none of these local development initiatives has a specific remit to target those who are socially excluded or in danger of becoming so. Pringle (1999) is also wary about local area-based approaches and argues that it is a mistake to assume that resources will be effectively directed to those in most need simply because programmes are administered locally and, in practice, it is those who are already well resourced who are best placed to tap into new sources of public funding. In short, where funding is the product of a bidding process the "Inverse Care" principle usually applies: i.e. the likelihood of an area receiving assistance is inversely related to its need for assistance. In such situations, funding tends to be allocated to those areas which make the strongest case (because they already have the capacity and resources to do so) while those which are most in need receive least because they lack the capacity to stake their claim effectively.

The area-based partnerships are still experimental and may become useful additions to the array of strategies available for tackling the problems that *accompany* marginalisation and social exclusion. However, the problems *also* need to addressed at a higher level. It is difficult to see how local, area-based solutions can eliminate the problems experienced in areas of multiple deprivation where the *causes* of those problems originate outside, and not within, the area. There is a danger that these new area-based approaches may turn out to be a cosmetic exercise that simply tackles the *symptoms* of the problems without addressing their real causes. It would be a tragedy if the impression is again created that something is being done if in reality these efforts simply serve to deflect attention from the macro-economic policies and the wider forces of global economic development which ultimately produce and compound the processes of social exclusion and marginalisation.

North Clondalkin has been included in those areas which are to receive funding under the European Union Urban Initiative (Taoiseach's Office, 1996). The area has also been included in the *Draft Area Action Plan* prepared by the Clondalkin Partnership for the wider Clondalkin area. Energy, enthusiasm and resources are being mobilised within the area and it will be interesting to see where this leads. In the meantime, North Clondalkin continues to be physically and socially isolated from, and invisible to, more prosperous parts of the city.

REFERENCES

Andersen, J. and Larsen, E. (1995) The Underclass Debate — A Spreading Disease? in Mortensen and Olofsson (eds.) *Essays on Integration and Marginalisation*, Samfundslitteratur, Copenhagen.

Bannon, M.J. (1989) Development Planning and the Neglect of the Critical Regional Dimension, in Bannon, M.J. (ed.) *Planning: The Irish Experience 1920-1988*, Wolfhound Press, Dublin.

Bartley, B. (1995) Managing Cities — Conflict Interpretations of City Organisation, in *Sociology: The Changing Social Environment,* OSCAIL, Dublin City University.

Bartley, B. (in press) Consuming Dublin — the Transforming City: Visibility and Invisibility — Spaces of Control and Excluded Places, in *Pleanail*.

Barrf, R. (1995) Multinational Corporations and the New International Division of Labour, in Johnston, R.J., Taylor, P.J. and Watts, M.J. (eds) *Geographies of Global Change*, Blackwell, Oxford.

Brotchie, J., Batty, M., Blakely, E., Hall, P. and Mewman, P. eds. (1995) *Cities in Competition: Productive and Sustainable Cities for the 21st Century*, Longman, Australia.

Clondalkin Partnership (1996) *Draft Area Action Plan: A Plan for the 21st Century*, Clondalkin Partnership, Dublin.

Collins, T. and Crowley, G. (1993) *Plan for the Development of North Clondalkin: Report of the North Clondalkin Task Force*, North Clondalkin Community Development Programme, Dublin.

Conlon, P. (1988) Public Transport -What is its Role? in Blackwell, J. and Convery, J. (eds.) *Revitalising Dublin — What Works?*, Resource and Environmental Policy Centre, University College Dublin, Dublin.

Cullingworth, J.B. and Nadin, V. (1994) *Town and Country Planning in Britain*, Routledge, London.

Davis, J. and Prendergast, T. (1995) Dublin, in Berry, J. and McGreal, S. (eds) *European Cities, Planning Systems and Property Markets*, Spon, London.

Davis, M. (1992) *City of Quartz; Excavating the Future in Los Angeles*, Vintage, New York.

Department of Justice (1992) *Urban Crime and Disorder: Report of the Interdepartmental Group*, Stationery Office, Dublin.

Dublin County Council (1983) *Dublin County Development Plan*, Irish Life Centre, Dublin.

Dublin Transportation Task Force (1986) *The East Regional Development Report — Review and Prospects*, Department of Transport, Dublin

Engels, F. (1969) *The Conditions of the Working Class in England*, Panther, England.

Fainstein, S. (1994) *The City Builders: Property, Politics and Planning in London and New York*, Blackwell, Oxford.

Gerckens, L.C. (1994) American Zoning and the Physical Isolation of Uses, *Planning Commissioners Journal*, 15.

Greed, C. (1993) *Introducing Town Planning*, Longman, Harlow.

Hall, P. (1989) *Urban and Regional Planning*, 3rd ed., Unwin, London.

Hartshorn, T.L. (1992) *Interpreting the City: An Urban Geography*, 2nd ed., Wiley, New York.

Harvey, D. (1989) *The Urban Experience*, Blackwell, Oxford.

Healey, P., Cameron, S., Davoudi, S., Graham, S., Madani-Pour, A. (eds.) *Managing Cities: The New Urban Context*, Wiley, Chichester.

Hutton, W. (1996) *The State We're In*, Vintage, London.

Jacobs, J. (1962) *The Death and Life of Great American* Cities, Jonathan Cape, London.

Kearns, G. and Philo, C.(eds) (1993) *Selling Places*, Pergamon, Oxford.

Kirk, G. (1980) *Urban Planning in a capitalist Society*, Croom Helm, London.

Knox, P.L. (1994) *Urbanisation: An Introduction to Urban Geography*, Prentice Hall, New Jersey.

Knox, P.L. (1995) *Urban Social Geography*, 3rd ed., Longman, London.

Lefebvre, H. (1974) *La Production De L'espace*, Anthropos, Paris.

Leyshon, A. (1995) Annihilating Space? The Speed-up of World Communications, in Allen, J. and Hamnett, C. (eds.) *A Shrinking World? Global Uneveness and Inequality*, Oxford University Press, Oxford.

MacLaran, A. (1993) *Dublin: The Shaping of a Capital*, Belhaven, London.

Massey, D. (1994) *Space, Place and Gender*, Polity Press, London.

McCarron, G. (1988) *County of Dublin Development Plan Review: Working paper No. 10 — The Neighbourhood Concept*, Dublin County Council Planning Department, Dublin.

McCormick, J. and Philo, C. (1995) "Where is Poverty? The Hidden Geography of Poverty in the United Kingdom", in Philo C (ed.) *Off the Map: The Social Geography of Poverty in the UK*, CPAG, London.

McGuirk, P. (1995) Power and Influence in Urban Planning: Community and Property Interests' in Dublin's Planning System, *Irish Geography*, 28 (1).

Mingione, E. ed. (1996) *Urban Poverty and the Underclass: A Reader*, Blackwell, Oxford.

Musterd, S. (1994) A Rising European Underclass? Social Polarization and Spatial Segregation in European Cities, *Built Environment*, 20(3).

Nexus (1996) *Partnership in Action*, Community Workers' Co-Operative, Dublin.

O'Toole, F. (1996) Walking a Fine Line Between Anarchy and Inertia, *Irish Times*, 6th September.

Pahl, R. (1997) Managers, Technical Experts and the State, in Harloe, M. (ed.) *Captive Cities*, John Wiley, London.

Painter, J. (1995) The Regulatory State: The Corporate Welfare State and Beyond, in Johnston, R.J., Taylor, P.J. and Watts, M.J. (eds) *Geographies of Global Change*, Blackwell, Oxford.

Peck, J. (1996) Loose Talk and Tight Fists, *The Guardian*, August 17th.

Prestwich, R. and Taylor, P. (1990) *Regional and Urban Policy in the United Kingdom*, Longman, London.

Pringle, D.G. (1999) Something Old, Something New: Lessons to be Learnt from Previous Strategies of Positive Territorial Discrimination, in Pringle, D.G., Walsh, J. and Hennessy, M. (eds) *Poor People — Poor Places: Poverty: Patterns, Processes and Policies*, Oak Tree Press, Dublin.

Room, G. (1993) *Anti-Poverty Action-Research in Europe*, School for Advanced Urban Studies, Bristol.

Sabel, C. (1996) *Local Partnerships and Social Innovation*, OECD, Paris.

Saunders, P. (1986) *Social Theory and the Urban Question*, 2nd edition, Routledge, UK.

Soja, E. (1996) *Thirdspace: Journies to Los Angeles and Other Real and Imagined Places*, Blackwell, Oxford.

Sibley, D. (1995) *Geographies of Exclusion: Society and Differences in the West*, Routledge, London.

Taoiseach's Office (1996) *Operational Programme: URBAN — Ireland*, Government Stationary Office, Dublin.

Walsh, J.A. (1999) Integration and Exclusion in Rural Ireland, in Pringle, D.G., Walsh, J. and Hennessy, M. (eds) *Poor People — Poor Places: Poverty Patterns, Processes and Policies*, Oak Tree Press, Dublin.

Ward, S.V. (1994) *Planning and Urban Change*, Paul Chapman, London.

Wilson, W.J. (1987) *The Truly Disadvantaged: The Inner City, The Underclass, and Public Policy*, Chicago, University of Chicago Press, Chicago.

Wilson, W.J. (1992) Another look at The Truly Disadvantaged, *Political Science Quarterly*. 106(4), 639-656.

Wright, M. (1967) *The Dublin Region: Advisory Regional Plan and Final Report*, Vols. 1 and 2, Government Stationery Office, Dublin.

Part 4

POLICY RESPONSES

12

SOMETHING OLD, SOMETHING NEW: LESSONS TO BE LEARNT FROM PREVIOUS STRATEGIES OF POSITIVE TERRITORIAL DISCRIMINATION

Dennis Pringle
Department of Geography
National University of Ireland, Maynooth

INTRODUCTION

Poverty, irrespective of how it is defined, tends to be spatially uneven — i.e. it is generally more concentrated in some areas than in others. This has given rise, from time to time, to policy responses based upon the principle of positive territorial discrimination — i.e. small areas of special need are designated to receive additional resources in an attempt to counteract the processes assumed to contribute to localised concentrations of poverty. Many of these policy responses implicitly incorporate bottom-up principles — i.e. by purposely designating fairly small areas, it is assumed that the implementation of the schemes can be adjusted to specific local needs and be informed by local public opinion. Such policies, it should be stressed, are designed to supplement, rather than to replace, universalist policies of welfare provision.

Small-area territorial strategies to tackle poverty and related problems are a comparatively recent innovation in Ireland, but they have a longer tradition in other countries. However, there is a tradition of positive territorial discrimination in the sphere of Irish regional economic planning. The Congested Districts Board was established to improve living conditions in eight western counties as far back as 1891. The CDB introduced measures to consolidate farm holdings, develop infrastructure, resettle people and develop fishing and other small industries in the designated areas until it was discontinued soon after the establishment of the state in 1923 (Orme, 1970). The regional dimension was revitalised in the 1950s with the establishment of An Foras Tionscal to administer grants in designated "undeveloped" areas in 1952, the establishment of Gaeltarra Eireann to promote manufacturing industry in Gaeltacht areas in 1958, and the establishment of the Shannon Free Airport Development Company to develop airport-related industries in 1959 (Walsh, 1989). However, although these schemes incorporated a territorial dimension, the designated areas were extensive in size, the objectives were primarily economic (e.g. job creation), and organisation was essentially "top-down".

This chapter will focus on positive territorial discrimination in social policy. It is argued that many of the policies introduced in Ireland in recent years, whilst innovative in the Irish context, are similar to many of the schemes introduced in other countries in the 1960s and 1970s. Given that there is a danger of resources and energies being wasted "reinventing the wheel", this chapter reviews some of the lessons which can be learnt from the British and American experiences in the 1960s and 1970s.

POSITIVE TERRITORIAL DISCRIMINATION IN THE UK AND US

Economic planning in Britain has been characterised by a sequence of Special Areas, Development Areas and Assisted Areas since at least the 1930s (Robson, 1988), whilst the genesis of a designated area approach in social policy can be traced back to the early 1940s (Edwards and Batley, 1978). However, the major blossoming of territorial strategies, in both Britain and the United States, came with the "rediscovery" of poverty in the 1960s. Before then, it had been assumed in Britain that the Welfare State, introduced in the late-1940s along the lines outlined in the Beveridge Report (1942), would gradually eliminate poverty by providing free education, free health services and benefits to the elderly, disabled, unemployed and other needy groups. However, by the mid-1960s, due to a fall in the real value of benefits (and wages), it became obvious that the universalist policies of the Welfare State by themselves were insufficient to eliminate poverty and that the number of people living below the poverty line was actually increasing (although at a much slower rate than it subsequently did in the 1980s). The British government responded in the late-1960s by establishing a variety of schemes for channelling additional resources into areas of special need: Educational Priority Areas (EPAs) were designated to receive extra primary school funding in an attempt to break the cycle of deprivation: General Improvement Areas (GIAs) were identified to tackle environmental problems; and, following the establishment of the Urban Aid Programme to counteract multiple deprivation, a number of Community Development Projects (CDPs) were set up in 1969 to explore how these problems might be tackled (see below) (Eyles, 1979). These schemes were followed in the early-1970s by other area-based schemes, such as Housing Action Areas (HAAs) and Comprehensive Community Programmes (CCPs).

Many of these schemes were influenced by similar developments which had taken place slightly earlier in the United States. The initial impetus for these was provided by the Ford Foundation's Grey Area Projects, set up to tackle the problem of educational deprivation in selected inner cities. The wide-ranging area-based approach envisaged in these projects was adopted by the President's Committee on Juvenile Delinquency and Youth Crime, which was established in 1961, and which culminated in the creation of 17 area-based projects under the Community Action Program by the end of 1964 (Marris and Rein, 1972).

The Community Action projects, it might be noted, incorporated elements of what we would refer to to-day as a "bottom-up" ethos: each project was given a high degree of autonomy; the objectives were to utilise, work through and revitalise the existing bureaucratic state agencies, rather than to compete with them; each project was to consult widely with the communities; and each incorporated a high degree of research and evaluation as part of a learning process. The project teams quickly recognised that poor communities did not simply lack money, or the will to improve their situation: they also lacked the political power to implement change. Many project leaders saw the need to mobilise the poor and encourage them to assert themselves as one of the first tasks of community action. This inevitably brought them into direct conflict with federal, state and city governments. Unfortunately, some communities became "over-assertive" and the Community

Action Program was phased out following widespread ghetto riots in 1967 (Edwards and Batley, 1978). The CAP was replaced by the Model Cities Program which, whilst similar in many respects, was more "top-down" and exercised much tighter government control.

The British schemes generally tended to be more "top-down" from the outset, but they became increasingly so as time passed. The Urban Programme was expanded, in terms of resources, in 1977, but it also became much narrower in focus: the earlier broad-based social objectives were displaced by narrower economic and environmental objectives (i.e. in effect, jobs and housing). In keeping with the spirit of the times, resources tended to be used to provide investment grants for the private sector, rather than be put into the community for self-improvement. The new ethos was epitomised by 11 Enterprise Zones set up by the Department of the Environment to counteract inner-city dereliction and industrial decline in 1980 (Norcliffe and Hoare, 1982).

The Republic of Ireland was largely bypassed by most of the earlier developments, although there were a few notable but small-scale exceptions, such as the Rutland Street project in Dublin. However, positive territorial discrimination has been introduced in more recent years. Designated Areas, analogous to the British Enterprise Zones, were identified in Dublin and the other major cities by the Department of the Environment in 1986. This was followed by a splurge of area-based schemes in the 1990s: 12 local Partnerships established in a pilot scheme to combat long-term unemployment under the Programme for Economic and Social Progress (PESP) in 1991 were supplemented by 28 Groups to receive support under the Global Grant funded by the European Regional Development Fund and European Social Fund under the aegis of Area Development Management Ltd. (ADM) in 1992. These in turn were designated for further support under the Operational Programme for Local Urban and Rural Development in 1994 (Haase *et al.*, 1996). 17 rural-based Groups were selected for funding under the LEADER Initiative in 1992 (Kearney *et al.*, 1994). This scheme was extended in 1996 under LEADER II. Other recent area-based strategies include the Third EU Anti-Poverty Programme, the Community Development Programme co-ordinated by the Department of Social Welfare and the Combat

Poverty Agency, and the NOW, Horizon and Euroform initiatives of the EU (Haase *et al.*, 1996).

These schemes are to be welcomed. However, there is a danger that, in our enthusiasm, we may replicate the same mistakes made in Britain and the United States in the 1960s and early-1970s. This chapter therefore reviews some of the problems which arose the "last time around".

There would appear to be at least three sets of arguments which could be made in favour of an area-based approach. I will discuss these, and some of the lessons which may be learnt, under three headings:

1. Targeting

2. Flexibility

3. Counteracting spatial processes.

TARGETING

One argument in favour of an area-based approach is that it provides a means for targeting additional resources at those who are most in need. However, this argument can only be sustained if it can be demonstrated that the areas which are designated are the areas of *greatest* need. The experience in Britain was that the areas of greatest need were not necessarily designated for preferential treatment. In fact many commentators actually suggested that there was an *"inverse care law"* in operation. The expression was first used in a medical context by a Welsh GP, John Tudor Hart, who argued that:

> In areas with most sickness and death, general practitioners have more work, larger lists, less hospital support and inherit more traditions of clinically ineffective consultation than in the healthiest areas; and hospital doctors shoulder heavier case-loads with less staff and equipment, more obsolete buildings and suffer recurrent crises in the availability of beds and replacement staff. These trends can be summed up as the inverse care law: that the availability of good medical care tends to vary inversely with the need of the population served (Hart, 1971, 412).

It was suggested, in a similar manner, that the likelihood of a deprived area being designated for special assistance was inversely related to its need for assistance.

There are several reasons why this should have been the case. Designation was not necessarily based on an objective assessment of need, but rather in many instances was the product of a bidding process in which community groups or local authorities applied for assistance. In such situations, funding tended to be allocated to the areas making the strongest case, rather than to the areas of greatest need. Electoral considerations also played a role and governments tended to favour moderately deprived areas in political marginal constituencies rather than the most deprived areas in "safe" Labour constituencies. In some schemes, local authorities had to provide matching funds — the problem here was that the most deprived cities did not have the resources to provide matching funds for all their deprived areas, whereas affluent cities with small pockets of deprivation could. A moderately deprived area in a prosperous area therefore had a greater chance of being designated than a severely deprived area in deprived city. Also, given the experimental nature of many of the schemes, there was a tendency amongst those advocating the schemes to select areas where there was a reasonable hope of producing a demonstrable improvement, in order to justify further funding for more difficult areas.

There would seem to be many parallels in the way in which areas were selected in the Irish area-based schemes. For example, ADM was restricted in its selection of Groups for support under the Global Grant to areas from which applications were received: if no application was received from a particularly disadvantaged area, then it could not be included in the Global Grant. Haase *et al.* (1996) suggest that the areas selected by ADM for the Global Grant correspond much more closely to real need, as objectively identified, than the areas selected for the PESP pilot scheme. Nevertheless, close examination of the maps in their report would subjectively suggest that there are many disadvantaged areas which did not receive support. It is also unclear as to what extent the *amount* of support received by each scheme was determined by need or by other considerations.

One alternative to allocating resources in response to the quality of applications received is to try to identify, using some objective method, which areas have the greatest need. This approach was widely adopted in Britain in the late-1960s and early-1970s, but these supposedly objective studies turned out to be much more subjective than intended (or realised). There is no simple objective index of poverty which may be used to compare and rank areas. Rather, one is forced to quantify poverty in an indirect manner, which introduces a high degree of subjectivity. There are two main reasons for using indirect measures:

- Neither the British nor the Irish census provides information on disposable income, whilst other sources of information on income tend to be either unreliable, unavailable or incomplete. It is therefore necessary to measure income indirectly using measures of things which are believed to be correlated with it (e.g. unemployment, education level, social class).

- It is widely accepted that poverty is a multidimensional phenomenon. Even if reliable measures of income were available, many would argue that they would need to be supplemented with measures of opportunity, powerlessness and other less-tangible attributes which condition the quality of life.

The need for multidimensional and indirect measures leads many researchers to adopt a methodology based on Factor Analysis or Principal Components Analysis. These are complex multivariate techniques which supposedly enable one to identify the underlying dimensions of variation within a large body of data. In other words, they may be used to condense the information on a large number of surrogate measures of poverty into a small number of indicators (or factors). Factor Analysis, however, gives rise to a large number of methodological problems (Pringle, 1978). There is no need to detail all of them here — one or two examples should suffice.

One of the best known problems is usually referred to as GIGO (Garbage In Garbage Out). In other words, if the analysis uses very poor surrogate measures of poverty, the composite measures will also be poor in quality. Unfortunately, the complexity of the techniques often blinds people to this simple truism. Another problem is that the relative importance and the definition of the factors identified is not determined by how pertinent they are as indicators of poverty, but rather by the extent to which the input variables are inter-correlated. For example, if one includes several measures of demographic structure in the analysis (simply because the data are readily available from the census), the chances are that they will tend to have comparatively high inter-correlations and will therefore tend to load high on the first principal component or factor. The composite measure of "poverty" will consequently tend to be more strongly influenced by demographic structure than by variables with lower correlations, but possibly having more direct relevance to poverty, such as unemployment. There is no intrinsic reason why poverty is necessarily related to high percentages of young people, or to high percentages of old people, yet these features may end up exerting a stronger influence on the composite poverty index than unemployment.

Factor scores are basically weighted sums of the composite variables, in which the weights are determined by the degree of inter-correlation between the variables. These weights, I would suggest, should ideally be based on some theory-based notion of the relative importance of the composite variables as measures of poverty, rather than being determined by a complex, but subjective, computer-based technique which few people really understand. One of the lessons of the 1970s was that factor analysis is no substitute for good theory (or even for bad theory).

There are many other technical problems associated with the methods used to quantify need, although they do not necessarily invalidate the usefulness of the exercise provided that the problems are recognised. A much more serious consideration, and more fundamental limitation, in the context of targeting is that, even if the most deprived areas are correctly identified and designated, area-based strategies provide a very inefficient means of targeting the people most in need, for the simple reason that the most deprived areas include only a relatively small percentage of the total number of people in need, and also include a large number of people who are not in need.

This point may be illustrated using data derived from a study of social deprivation in Belfast in the early 1970s (Boal et al., 1974; 1978). Belfast was divided for the purposes of this study into 93 sub-areas. Table 12.1 illustrates the effectiveness of designating the 10, 20 and 30 most deprived areas for preferential treatment. For example, the 10 worst areas include 9.88 per cent of all households, and 18.76 per cent of the adult male unemployed. In other words designating these 10 areas, despite including almost 10 per cent of the entire population, would have excluded more than four fifths of the unemployed. An even higher percentage would be excluded if one required the 10 designated areas to be spatially contiguous. To include even one half of the total unemployed in the designated area, one would have to designate an area containing almost one third of the entire population. The unemployed would represent a comparatively small percentage of the total population living in the area. The situation would be much the same with regard to trying to target assistance for other types of deprivation. Table 12.1 shows the percentages which fall within the 10, 20 and 30 worst areas for various other indicators of need. Area-based strategies, in short, do not provide a very efficient means for targeting people in need, even in a highly socially-segregated city such as Belfast.

Table 12.1: Percentage of Different Types of Social Problem Which Would be Enclosed in Designated Areas of Different Sizes in Belfast

	Area	**Households**	**Unemployed**	**Children in Care**	**Juvenile Offenders**	**Infant Mortality**
10 worst areas	3.60	9.88	18.76	26.62	18.70	13.88
20 worst areas	7.79	19.84	32.64	47.75	32.30	25.98
30 worst areas	13.57	31.61	51.28	64.65	52.85	43.06

Area-based strategies can probably be assumed to be even less efficient in cities with lower levels of socio-spatial segregation, and almost totally ineffective in rural areas where rich and poor tend to be spatially more interspersed. In fact, the concept of targeting is probably almost meaningless in a rural context, so it is arguable whether there is any point in trying to identify which *areas* are the most deprived. The 17 LEADER projects, for example, covered 61 per cent of the total area of the state (Kearney *et al.*, 1994), so targeting (at least in a spatial sense) was presumably not a major consideration.

FLEXIBILITY

A second set of arguments in favour of area-based strategies, irrespective of their effectiveness as a targeting mechanism, is that they provide the potential for greater flexibility in service delivery. The US Community Action projects in the 1960s tended to centre on a number of recurrent themes, which can be summarised by the words: *co-ordination, innovation* and *participation* (Marris and Rein, 1972). There are obvious parallels in the philosophy underlying the local development approach promoted by the Global Grant which may be summarised by the concepts: partnership, participation, planning and multidimensionality (Haase, *et al.*, 1996).

The Community Action projects recognised that poverty was a multi-dimensional phenomena, the causes and symptoms of which fell under the jurisdiction of the different government departments responsible for employment, housing, education, crime and delinquency, etc. Each department tended to have a different outlook and tended to operate with minimal consultation with other departments, resulting in unnecessary duplication between departments and gaps in coverage where specific problems fell into the jurisdictional cracks. Area-based strategies seemed to provide a means whereby the activities of the various government departments could be more effectively co-ordinated. By dealing with smaller areas, it was hoped they would provide the flexibility to design programmes which could be accommodated to the specific needs of particular areas. It was also hoped that they would provide greater flexibility to adjust in response to changing circumstances. It was envisaged that the projects would work through the government departments, rather than compete with them.

The Community Action projects recognised that the causes, and hence the solutions, to poverty were poorly understood. Government departments charged with responsibility for addressing these problems were, like all large bureaucracies, characterised by a high degree of inertia. The individual projects were therefore encouraged to experiment and to be innovative in the ways in which they tackled the problems. Most of the schemes in the US and UK consequently incorporated a large element of research and evaluation.

The third key concept was "participation". It was recognised that local communities were usually excluded from the decision making process, yet they contained a wealth of information and expertise. One of the principal goals of the Community Action projects was to try to tap into this unused potential by utilising what we would refer to today as a "bottom-up" approach.

These remain admirable objectives and, in my opinion, still provide very persuasive arguments in favour of area-based strategies. However, the implementation of these objectives in practice proved much more problematic than had originally been envisaged. Although area-based schemes were intended to facilitate greater co-ordination between government departments, different departments in practice tended to act defensively to safeguard their own particular spheres of influence. Many of the projects were also much less innovative than had originally been intended — given limited resources, there was a tendency for projects to borrow ideas which appeared to work in other areas, rather than implement a more imaginative response of their own. Indeed, many of the projects ended up replicating services provided by government departments, which not only intensified problems of duplication but antagonised relations between the local-based project leaders and the government departments with whom they should have been co-operating.

Most of the problems, however, related to the bottom-up aspects of the projects. Terms such as "top-down" and "bottom-up" tend to suggest a two-tier structure comprising a "top" (i.e. government) and a "bottom" (i.e. local communities). In practice, the projects tended to be administered by project teams (including academics) who constituted a third tier, intermediate between top and bottom. This third tier can exert a powerful controlling influence, with the result that what appears as "bottom-up" to the top (i.e. government) may appear as "top-down" to the bottom (i.e. local communities). In many instances, the situation was further complicated by conflicting interest groups within each tier. All sorts of unanticipated stresses emerged in the Community Action projects: jurisdictional disputes developed between project leaders and government departments; local politicians resented the influence that non-elected project leaders had in determining local policies; and project leaders tended to interpret what was meant by "participation" in different ways. Some project leaders interpreted participation as meaning little more than "consultation" with the target communities, resulting in what might be described as a "localised top-down" approach, in which the local communities were effectively frozen out of the decision making process (or else only brought in when it was too late to have much impact). In other instances, radical project leaders mobilised and politicised the community, resulting in an orchestrated protest by communities against government. Although the analyses of the causes of the problems by these "hot-headed" project leaders were often correct, the tactics adopted proved to be self-defeating because the government retaliated by simply discontinuing the funding of projects.

The precise nature of the problems which emerged tended to vary considerably depending upon the specific local context. Nevertheless, the net effect was that the original concept of participation in decision-making tended to be watered down and replaced by more restricted mechanisms (e.g. "consultation"), which of course is more amenable to top-down management.

COUNTERACTING SPATIAL PROCESSES

The fact that poverty tends to be more concentrated in some areas than in other areas gives rise to a third set of arguments in favour of area-based strategies. Common sense suggests that these spatial concentrations of poverty must be the product of geographical processes: if these processes can be identified, it may be possible to counteract them at local level.

It is important, in this regard, to make a distinction between processes operating within a local area which increase the likelihood of poverty developing within that area; and processes which increase the likelihood of poverty, which may have its origins elsewhere, becoming concentrated in certain areas. Examples of the first, in a rural context, might include: poor land quality, a past history of excessive land subdivision, poor infrastructure, or poor accessibility to the nearest town. Each of these factors is likely to impair the life chances of people living within that area. Examples of the second type of process would include migration preferences and, in an urban context, the filtering effects of the housing market. People on low incomes tend to be excluded from the owner occupied housing sector because of their inability to obtain a mortgage; they consequently tend to end up in either the public or private rented sectors which are spatially concentrated in certain parts of cities. If the public housing authority operates a policy of allocating the "problem families" (e.g. rent defaulters) to the least desirable public housing stock, poverty will tend to be even more spatially concentrated. The poor, in other words, will tend to be concentrated in certain areas, not because the areas themselves "cause" poverty, but because the poor are in effect excluded from other areas. Spatial concentrations of poverty are not caused solely by processes operating within deprived areas; they are also caused by broader structural processes which operate within society to create social inequalities and then operate to sort those who are least able to compete into the less desirable (i.e. deprived) areas by excluding them from the more affluent areas. (Incidentally, one should not overlook the important sub-category of "middle class poor" — people who own their own houses, but who spend so much of their disposable income on mortgage repayments that they have little available for anything else. Many people in this category tend to end up in the lower end of the owner occupied housing sector, which again tends to be spatially concentrated in certain areas).

Area-based strategies provide a possible mechanism for dealing with the first set of processes (i.e. local processes which increase the likelihood of poverty). However, they are relatively ineffectual as a means for tackling the second (i.e. broader structural processes). It is therefore necessary to accurately diagnose the processes which give rise to a concentration of poverty in a particular area.

The British Community Development Projects (CDPs) are particularly instructive in this regard. These were established in 1968 at a time when concepts such as the "cycle of deprivation" were fashionable. The prevailing assumption before the CDPs was that:

> The welfare state could solve people's individual problems, but when a significant number of those same people were concentrated in geographical areas — the old, the unskilled, the disabled, the unemployed left behind by the tide of industrial change — they became "multiply deprived". Whole families became caught in a "cycle of deprivation" that was not only "transmitted from generation to generation" like some hereditary disease but was also immune to the widely canvassed cure of "equal opportunity" (CDP, 1977, 54).

The objective of the CDPs was to identify how people in deprived areas could be helped to solve their own problems. However, the CDPs, working in 12 different areas, gradually came to the conclusion that the fault lay not with the poor, but with the structures of society:

> ... this poverty is a reflection of inequalities in society as a whole. Clearly the scale and character of the problem is too great for policies concerned solely and specifically with inner (city) areas to be effective. Any fundamental change must come through policies concerned with the distribution of wealth and the allocation of resources (CDP, 1977, p. 24).

After further analysis, the CDP teams came to the conclusion that the British area-based schemes owed much more to government concerns about crisis management than to a concern for the plight of the poor. By identifying small areas of "multiple deprivation", the problems of poverty and inequality within British society could be depicted as being confined to a few small black spots and therefore dismissed as being in some way abnormal (Carney and Taylor, 1974). Also, by launching high profile, but low cost, programmes to tackle these problems, the government could create the impression of trying to do something about the problem, whilst in practice doing very little. The whole area-based strategy consequently came to be dismissed by those most actively involved as an exercise in "cosmetic planning" (e.g. Duncan, 1974).

This critique naturally brought the project teams into conflict with their paymasters in the Home Office, and the projects were eventually closed down in 1976. Their reports were suppressed by the government, but not before a Home Office spokesperson admitted that:

> "Gilding the ghetto" or buying time, was clearly a component in the planning of both CDP and Model Cities (the US Poverty Programme) (CDP, 1977, p. 46).

Conclusion

Area-based strategies to combat poverty and deprivation are by no means new. Past experience suggests that they are at best an inefficient method for targeting disadvantaged groups for preferential treatment. However, if areas are to be selected for preferential treatment, then it would seem desirable that the areas selected should at least be those most in need. These selections should be based upon an objective nation-wide study of need. Such studies are methodologically problematical, but most of the problems are well documented. The most immediate requirement is to develop a composite index (or indices) of poverty, suitable for the Irish context, in which the weights assigned to the component parts are based on a theory of relative importance rather than determined by complex, but arbitrary, statistical techniques.

One of the major strengths of an area-based strategy is the potential flexibility it provides for innovation. Area-based strategies should attempt to maximise the resources, drive and knowledge which exists within communities. This would require communities to be adequately empowered and resourced, but past experience suggests that there are many problems to be overcome, even assuming that the political will for more open and democratic local government existed. Contemporary area-based strategies tend to be geared towards local economic development (i.e. job provision). Job creation is obviously important in the war against poverty, but area-based approaches provide the potential for a much broader-based and more integrated strategy. Much of the innovative idealism of the 1960s was probably misplaced, but we seem to have swung too far in favour of narrow pragmatic objectives.

Area-based strategies have the potential for counteracting processes which help to create poverty within disadvantaged areas. However, it must be recognised that many of the causes of poverty do not originate at local level, nor do local spatial concentrations of poverty necessarily indicate local causal processes. If local area-based strategies are to be effective it is essential that we fully understand the nature of the causal processes. Action-based projects therefore need to be supplemented with objective assessments of the causal processes. The alternative is that, even with the best of intentions, we may end up inadvertently "gilding the ghetto" once again.

REFERENCES

Boal, F.W, Doherty, P. and Pringle, D.G. (1974) *The Spatial Distribution Of Some Social Problems In The Belfast Urban Area*, Northern Ireland Community Relations Commission, Belfast.

Boal, F.W, Doherty, P. and Pringle, D.G. (1978) *Social Problems In The Belfast Urban Area: An Exploratory Analysis*, Occasional Paper No. 12, Department of Geography, Queen Mary College, London.

Carney, J.G and Taylor, C. (1974) "Community Development Projects: review and comment", *Area*, 6(3), 226-231.

CDP (1977) *Gilding the Ghetto. The State and the Poverty Experiments*, CDP Publications, Newcastle.

Duncan, S.S. (1974) "Cosmetic planning or social engineering? Improvement grants and improvement areas in Huddersfield", *Area*, 6(4), 259-271.

Edwards, J. and Batley, R. (1978) *The Politics of Positive Discrimination. An Evaluation of the Urban Programme 1967-77*, Tavistock, London.

Eyles, J. (1979) "Area based policies for the inner city: context, problems, and prospects", in Herbert, D.T. and Smith, D.M. (eds.) *Social Problems And The City*, Oxford University Press, London.

Haase, T., McKeown, K. and Rourke, S. (1996) *Local Development Strategies for Disadvantaged Areas. Evaluation of the Global Grant in Ireland (1992-1995)*, Area Development Management, Dublin.

Hart, J.T. (1971) "The inverse care law", *Lancet*, 1, 405-12.

Kearney, B., Boyle, G.E. and Walsh, J.A. (1994) *EU LEADER I Initiative in Ireland. Evaluation and Recommendations*, Department of Agriculture, Food and Forestry, Dublin.

Marris, P. and Rein, M. (1972) *Dilemmas of Social Reform*, 2nd ed., Routledge and Kegan Paul, London.

Norcliffe, G.B. and Hoare, A.G. (1982) "Enterprise Zone Policy for the Inner City", *Area*, 14(4), 265-276.

Orme, A.R. (1970) *Ireland*, Longman, London.

Pringle, D.G. (1978) *Spatial Analysis of Urban Social Problems: Some Methodological Considerations*, Unpublished Ph.D. thesis, Department of Geography, Queen's University, Belfast.

Robson, B. (1988) *Those Inner Cities. Reconciling the Social and Economic Aims of Urban Policy*, Clarendon, Oxford.

Walsh, J.A. (1989) "Regional development strategies", in Carter, R.W.G. and Parker, A.J. (eds.) *Ireland. Contemporary Perspectives on a Land and Its People*, Routledge, London.

13

THE ROLE OF AREA-BASED PROGRAMMES IN TACKLING POVERTY

Jim Walsh
Combat Poverty Agency, Dublin

INTRODUCTION

Since the early 1990s, there has been a growing emphasis in welfare policy on spatial programmes to tackle unemployment, poverty and social exclusion. Following a number of pilot schemes, area-based programmes are now firmly part of government policy for tackling poverty. This is most clearly evidenced in the National Anti-Poverty Strategy (Ireland, 1997), which exhibits an explicit spatial dimension in two of its five priority themes: disadvantaged urban areas and marginalised rural communities. This spatial focus to welfare policy is further reflected in a variety of spatial anti-poverty programmes, most notably the Programme of Integrated Development in Designated Areas of Disadvantage, part of the EU-funded Local Development Programme. This area focus in welfare policy is influenced by the contemporary geography of poverty, which reveals that 50 per cent of poor households are to be found in public housing, with 30 per cent in the estates of the five largest local authorities (Nolan *et al.*, 1998). It also derives from a decentralisation of welfare policy, prompted by the desire to improve local delivery, to engage with local communities and to promote local employment initiatives.

There is a considerable literature on the content and impact of area-based programmes: e.g. Cullen (1989; 1994); Craig and McKeown (1994); Harvey (1994); Walsh (1994a); and Haase *et al.* (1995). Meanwhile, NESC (1994), OECD (1996), Walsh *et al.*

(1997; 1998) highlight their institutional and policy implications. Despite this rich literature, there has been little attempt to critique the spatial dimension in these programmes. This is not to deny that some debate has taken place on the efficacy of various poverty indices and the degree to which poor households are concentrated in specific localities (Nolan *et al.*, 1994; 1998; Haase *et al.*, 1996). However, wider geographical issues have been almost completely ignored, such as the processes generating spatial inequalities (e.g. housing market, agri-business, financial services) and the living conditions in poor areas (e.g. access to services, quality of housing and the environment, communal resources). An understanding of these issues should then influence the content, delivery and scale of area programmes. This chapter will review how these geographical issues have informed and been addressed by a variety of spatial poverty programmes.

Interestingly, the Irish policy preoccupation with spatial programmes is not without precedent. Pringle, in this volume, recalls a lively debate in the 1960s and 1970s about spatial approaches to social problems in North America and the UK. This chapter seeks to bridge the gap between the experiments of the past and the enthusiasm of the present. It begins by tracing the emergence of area-based programmes as a policy instrument for tackling poverty. It then reviews the application of the spatial dimension in the current proliferation of such programmes using a threefold classification. The chapter concludes by suggesting ways in which geographical issues might be better addressed both in specific poverty programmes and in welfare policy in general.

LOCALISM IN WELFARE POLICY: THE EMERGENCE OF AREA-BASED POVERTY PROGRAMMES

A spatial dimension to welfare policy is not a new departure. The poor law in the 19th century was administered on the basis of a local "union" (a collection of townlands) (Powell, 1992). Under this localised structure, an elected board of guardians funded and administered relief to those considered in need in these unions. Another 19th century local welfare initiative was the Congested Districts Board, a government body which sought to address poverty and over-population in designated parts of the west of Ireland. Its actions included improvements in agriculture and fishing, population transfer, land reform and home management training. Since then, the spatial dimension of welfare policy has greatly declined, and now only exists as a residual feature in administrative schemes such as free school meals (confined to the Gaeltacht and urban areas) and Supplementary Welfare Allowance (with its localised delivery known as the community welfare service). This decline is in keeping with the emergence of a centralised welfare system.

A residual spatial dimension is also evident in other realms of public policy. Industrial policy has for a long-time featured a differentiation in the level of grant aid payable between the east and west of the country. Similarly, in agriculture, compensation payments are made to farmers in areas classified as "handicapped" due to difficult land conditions. Other examples include the additional provision for Gaeltacht areas and the designation of run-down urban areas, and more recently certain seaside resorts and rural areas, for special tax relief on private investment in housing and commercial development. In all these instances, the spatial dimension is used as a crude administrative tool of central government to target resources. In only a minority of cases is this resource ring-fencing accompanied by a devolution of administrative powers, such as some regional development and urban regeneration agencies, e.g. Udaras na Gaeltachta, Dublin Docklands Development Authority. Again, this situation highlights the dominance of central government in Irish public administration, including the allocation of EU structural funds.

In was against this background that a significant shift in public policy took place from the late 1980s, which led to the incorporation of a spatial dimension in strategies for tackling poverty. This new-found geographic emphasis was due to a combination of factors, which are described below.

A New Geography of Poverty
The growth in poverty that occurred in the 1980s (Callan *et al.*, 1996) was accompanied by a new geography of poverty. Rising poverty levels had a distinct urban aspect, stemming from high rates of unemployment in local authority housing estates. These estates suffered both from the general increase in unemployment

in the 1980s, especially among unskilled manual workers, and the targeted exodus of employed people from these estates under housing surrender grants. The resultant high rates of joblessness and poverty were described in numerous local profiles, including the CODANS (SUS Research, 1987) study of county Dublin, giving rise to the popular term of unemployment "blackspots". Interestingly, this analysis pre-dated studies on the overall distribution of unemployed households (Combat Poverty Agency, 1993; Nolan *et al.*, 1994). Nonetheless, the concept of unemployment blackspots quickly assumed a policy importance, with the concentrated incidence of unemployment and poverty in certain communities being used to provide a "prime facie case for the development of area-based programmes" (NESC, 1990, p72). This view was further emphasised in the National Anti-Poverty Strategy, where it states:

> As well as considering the three core issues of educational disadvantage, unemployment and income adequacy, it is also necessary to examine the consequences of high levels and concentrations of poverty which can lead to a threat to the social fabric of the country and incur high economic costs. . . . There are areas throughout Ireland where there are concentrations of people living in poverty, often resulting in cumulative disadvantage. These are sometimes referred to as "poverty blackspots". It has been suggested that the experience of being poor and living in such areas is a qualitatively different, and usually worse, experience that being poor and living in a non-disadvantaged environment (Ireland, 1997, 16).

This analysis is given a clear policy expression through the advocacy of special measures to address urban and rural poverty concentrations. A widely perceived spatial pattern in the distribution of other, socially vulnerable categories, such as travellers, drug addicts, lone parents and educational disadvantaged children, has also been important as a prompt in policy circles for the introduction of locally targeted interventions.

Improving the Delivery of Welfare Policy

There were also significant developments within mainstream welfare policy that encouraged a greater local focus to the design and delivery of services. First, there was a growing desire among policy-makers for better targeting of resources on those in need, as part of a general move away from universalism in social policy. This case for greater efficiency was particularly strong in the late 1980s, when the fiscal crisis was at its height and severe cutbacks in public expenditure were required. In this context, spatial programmes provided an innovative means of identifying and responding to the emerging social needs of this period (NESC, 1990).

A second policy concern was to make service provision more effective through better local co-ordination, especially in response to the multi-dimensional nature of many intractable social problems, as indicated by the newly-coined phrase "social exclusion". The contribution of NESC in endorsing this perception was especially significant, with its 1990 strategic policy document offering the following critique of prevailing policy:

> Currently, social policies and services operate on a "functional" or "departmental" basic (health, social welfare, and others) without any coherent attempt to integrate services at local levels. Clearly, many low income communities are affected by the services, and receive resources from a range of state agencies — local government, health boards, the Department of Social Welfare, FAS, for example. The scope for area "renewal" and community based co-ordination must therefore be considerable. Evidence suggests that concerted, intensive programmes in small areas, containing elements of housing and environmental improvement, as well as retraining and employment schemes and "outreach" health and educational projects, can have an impact over and above the separate effects of individual programmes. Furthermore, the more closely involved are local communities in the planning and delivery of area-based projects, the more they will reflect local needs and priorities (NESC, 1990, 74).

Similar demands for greater integration of welfare services were enunciated in other official policy reviews, e.g. Interdepartmental Group on Urban Crime and Disorder, 1992; NESF, 1995; Task Force on Long-Term Unemployment, 1995; Kelleghan *et al.*, 1995; Ministerial Task Force on Measures to Reduce the Demand for Drugs, 1996. It was also a key theme in EU social policy, most notably in the Third EU Poverty Programme, where the multidimensional nature of poverty, as suggested in the term social exclusion, was to the fore.

Community Development as a Component of Welfare Policy
A further theme emanating from reform of welfare policy, alluded to in the NESC quote above, was a recognition of the contribution local "communities" could make to the planning and delivery of services. This was reflected in a growing reliance on community-based models of welfare provision, including information and advice (community information centres), labour market programmes (community training workshops, community employment programmes) and social services (community care) (McCashin, 1990; O Cinneide and Walsh, 1990). It was also reflected in the approaches of non-governmental bodies such as childcare organisations (family centres) and trade unions (centres for the unemployed). This focus on community has had a long undercurrent in welfare policy, based on a tradition of local self-help, e.g. Muintir na Tire, credit unions, though hitherto largely ignored in policy circles. The approach is usually referred to as community development, which extends beyond a notion of community as simply being a site for the delivery of public services, to community as having a pro-active role in deciding what and how services are provided.

This enhanced community role is particularly evident in the realm of anti-poverty policy. Here, community development explicitly refers to the empowerment of individuals and groups experiencing poverty and social exclusion. This understanding of community development is associated with the work of the Combat Poverty Agency from the mid-1980s. The Agency, a statutory body, promoted community development as a strategy for involving disadvantaged groups in deprived urban and rural areas in local decision-making, with the support of pilot initiatives such as the second EU poverty programme (Cullen, 1989; Combat Poverty Agency and Community Workers Co-operative, 1990). The Agency advocated that involving disadvantaged people in the process of change would, in turn, lead to gains in terms of the effectiveness of service provision (Combat Poverty Agency, 1989).

Over time, community development has also become evident in other areas of welfare policy, such as in education ("home-school-community liaison"), social housing ("tenant participation in estate management") and local economic development ("community capacity-building"). Meanwhile, at a national level, the community sector has received national recognition through the establishment of umbrella networks such as the Community Workers Co-operative, the Irish National Organisation of the Unemployed and the Irish Travellers Movement. These networks have been included in official policy bodies such as the National Economic and Social Forum and were a party to the social partner discussions leading to Partnership 2000. This legitimisation of the community sector as an actor in welfare policy (Curtin and Varley, 1995) is given explicit government expression in the recent green paper on voluntary activity (Department of Social Welfare, 1997).

Local Development and Employment Initiatives

The rise of local anti-poverty interventions can also be traced to influences emanating from economic policy. From the mid-1980s, there was a growing awareness of the potential of local development and employment initiatives as a new source of jobs in both national and EU economic policy. Again, a state agency, in this case the Youth Employment Agency, was crucial in giving official support to local employment initiatives through the Community Enterprise Programme (O Cinneide, 1985), while in Gaeltacht areas, some limited government support was provided for community co-operatives. Another government initiative in the late 1980s was the Area Programme for Integrated Rural Development, which fostering support for local economic development in a rural context. Subsequently, the policy merits of a "bottom-up" model of economic development in fostering entrepreneurship, redistributing economic activity and facilitating local involvement in the development process were articulated in an influential government-commissioned assessment of the 1989-93 Community

Support Framework (ESRI *et al.*, 1993). Meanwhile, at the international level, various EU policy reports in the early 1990s highlighted the importance of local development and employment initiatives in promoting job creation (Commission of the European Communities, 1993; 1995). This emphasis was reflected in the structural funds and innovative programmes such as Leader, Leda and Employment.

The emergence of local economic development had a knock-on effect on welfare policy. This occurred through their shared concern with unemployment as a policy issue, along with their common emphasis on community self-help as a way of delivering policy. Local development initiatives were seen as providing a means of integrating enterprise with welfare and economic growth with redistribution. Thus, local job creation in new sectors of the economy was a practical way of improving access to the labour market for disadvantaged groups. This new welfare/enterprise nexus is given strongest expression in the Local Development Programme, part of the Community Support Framework. This combines support for micro-enterprise, education and training measures and community development initiatives in a framework of integrated local development targeted at disadvantaged urban and rural areas. A similar alignment is evident in the Back-to-Work Allowance, a scheme that allows the unemployed to retain welfare benefits while establishing micro-businesses.

The remainder of this chapter will analyse a section of the various local anti-poverty initiatives that have been developed in recent years. These are categorised into three main themes:

- Targeting additional resources at areas of high social need;

- Enhancing local co-ordination and delivery of services in areas of multiple disadvantage;

- Promoting integrated local development in areas of cumulative disadvantage and underdevelopment.

These categories are not mutually exclusive and individual programmes may incorporate more than one theme. Also, these categories represent a range of sophistication, with resource targeting representing the crudest use of space and integrated development initiatives the most ambitious.

TARGETING OF RESOURCES

The first category of spatial anti-poverty programmes targets additional resources at designated areas of need. They operate on the assumption that by re-distributing resources to disadvantaged areas, the well-being of inhabitants can be enhanced relative to the norm. The two main examples of positive spatial discrimination relate to educational disadvantage and long-term unemployment. Other area programmes, which will be considered later on, also place an emphasis on re-distributing resources to areas of disadvantage, though only as a backdrop to more critical interventions such as improved service delivery, community development or local job creation.

Case Studies: Educational Disadvantage and Long-term Unemployment

In education, the Department of Education and Science has devised two schemes to target additional resources to schools serving areas of disadvantage. These provide higher capitation grants for general management, books and materials, home-school liaison schemes, and extra teaching and other specialist personnel. The two main schemes are the Scheme of Assistance to Schools in Designated Areas of Disadvantage and the Breaking the Cycle initiative. The Scheme of Assistance dates from 1990, although an *ad hoc* arrangement has existed since 1984. In 1996, 310 primary schools with 82,000 pupils were covered by the scheme, representing 9.5 per cent of schools and 16.7 per cent of pupils. The corresponding figures for second-level were 190 schools with 100,000 pupils, which represents 22.5 per cent of schools and 24 per cent of pupils. The cost of the scheme for primary schools in 1996 was £2.6m, plus the salaries of 293 additional teachers (costing around £6m) and, at second-level, £1.4m and 209 teaching posts (costing around £4.2m). Breaking the Cycle is a more intensive and targeted version of the general scheme, and is confined to primary schools in urban areas and small rural primary schools. Again in 1996, there were 156 schools in the scheme, with 14,300 pupils, at a total cost (grants and salaries) of £3m.[1] The main aspects of this initiative are local co-ordination, smaller class sizes, higher capitation payments, in-service training for teachers, and a five-year plan of action.

The main example of a spatially targeted employment programme is the Area Based Response to Long-Term Unemployment (ABR) and its linked programme, the Global Grant for Local Development (GGLD). These two initiatives operated on a pilot basis between 1991 and 1995, before being subsumed into the Local Development Programme. The ABR arose from the joint government/social partner Programme for Economic and Social Progress. Its aim was to tackle long-term unemployment in 12 pilot areas, 8 of which were urban. The core resources of the ABR (£4m) were topped-up by an EU global grant (£8m), though some of the latter (about 40 per cent) was made available to non-designated areas. The central aim of the ABR/GGLD was to bring about "an explicit and targeted redistribution of job chances" towards the long-term unemployed in spatially designed areas. To this end, the initiative provided additional resources for the purposes of retraining, enterprise support and job placement in the 12 localities, with the specific aim of "ensuring that one area (was) favoured over another" (Craig and McKeown, 1994, 27). The combined resource base of the programme (about £8m) was supplemented by additional resources, estimated at £65m, drawn mostly from statutory service providers, though not all of this would have been new monies (Craig and McKeown, 1994).

Limited Efficiency and Effectiveness of Targeting

Area programmes of positive discrimination have an understandable appeal in terms of channelling scarce resources to those most in need. In addition, such programmes fit easily with existing structures and thus have minimal operating costs. However, there are some weaknesses that impact on the efficiency and effectiveness of spatial targeting, which are now considered. In the first instance, how well area programmes are targeted depends on the accuracy and reliability of the indicators being used. This is part of a bigger debate about poverty indicators, which Haase and Williams *et al.* address earlier in this book. Kelleghan *et al.* (1995) raise serious doubts about the validity and appropriateness of the indicators of educational disadvantage used by the Department of Education and Science, identifying six substantial defects, e.g. validity of returns, indicator bias, arbitrary weighting. A second factor is the extent to which educational disadvantage and long-term unemployment are spatially concentrated. Here a key distinction must be made between risk — the percentage of people in poverty — and incidence — the overall distribution of poor people. A recent review of the Scheme of Assistance shows that pupils in designated primary schools have double the risk of educational disadvantage in general: 32 per cent compared to 16 per cent (Kelleghan et al., 1995). At the same time, such schools only serve a minority of all those who are educationally disadvantaged, estimated to be less than one-in-three (30 per cent). This share is likely to be somewhat higher at second-level because disadvantage is more concentrated in post-primary schools, but also due to the fact that designated post-primary schools are larger and thus account for a higher share of all pupils. The level of targeting achieved under the ABR can be questioned to an even greater basis. To begin with, some of the areas selected did not rank amongst those worst off, though this criticism has been rejected on the basis that the ABR was a pilot initiative.

Other distortions arise from the geographical unit used to delineate disadvantage. For example, the basic unit of measurement under the remedial educational programmes, the school, is influenced by size and catchment factors. Consequently, primary pupils in urban areas are more likely to be in a designated school (40 per cent) as compared to towns and rural area (14 and 5 per cent respectively), because of more segregated housing in urban areas and, secondly, larger schools serving such areas. Yet, educational disadvantage is primarily a rural phenomenon, accounting for 61 per cent of all disadvantaged pupils. As a result, provision to tackle educational disadvantage is 20 times more per disadvantaged pupil in Dublin than in rural areas (Kelleghan et al., 1995). Meanwhile, under the ABR and its successor programmes, attempts have been made to modify the influence of space by relying on census-based units (DEDs) to draw boundaries. However, these too are prone to bias due to variations in size and population mix (see Cooke et al. in this volume). They are further compromised by an operational desire to work with meaningful administrative units rather than artificial areas indicated by lines on a map.

We need also to consider whether the level of resources provided under these initiatives is sufficient to address the scale of disadvantage encountered by residents of these areas. An important starting point here is whether and how an adequate level of resources is arrived at. The evidence would suggest that this is not based on an objective measure of what is needed to overcome the current deficit, but on the government resources made available for targeting, which is essentially a budgetary decision. As noted in a review of the Scheme of Assistance:

> The extra assistance to schools just about brought their resources up to the level of schools serving non-disadvantaged areas. While the scheme has served a useful function in providing much needed resources . . . it cannot really be said to have operated positive discrimination in favour of designated schools. Given this situation, it would be unrealistic to expect it to have had much impact on students' achievements or life chances (Kelleghan *et al.*, 1995, 63).

In the instance of the ABR, vagueness as to the additional resources mobilised makes it difficult to measure the scale of redistribution. Even when more resources are made available from central government, as occurred over time with the Scheme of Assistance or through the expansion of the ABR through successor programmes, the potential impact is frequently diluted by a parallel expansion in the number of designated areas.

Equally important as the level of additional resources is the use made of these resources. In most cases, the additional funds are used to increase provision, be it in terms of additional personnel, administrative resources, or services. This emphasis on *quantity* leads to a neglect of issues such as the *quality* of education and training provided. The supply-side focus on the pupil or unemployed person also ignores system factors which may perpetuate disadvantage, e.g. the practice of inter-school selectivity resulting in "dump schools" for the least able students, the lack of demand for low skilled labour or the discriminatory recruitment practices of employers. The demand issue is also important given the redistributive nature of these programmes in regard to people's life chances, including a potentially worsening of the situation for disadvantaged individuals in non-designated areas.

In response to concerns about the efficiency and effectiveness of "top-up" schemes in targeting educational disadvantage, the Department of Education and Science introduced the Breaking the Cycle initiative. This 1996 initiative has two significant refinements: one is to concentrate on schools in urban areas with the highest rates of educational disadvantage; and the second is to focus on clusters of small (less than five teachers) rural schools. The overall outcome of this is confusing. In terms of the distribution of disadvantage, Breaking the Cycle creates a more balanced rural-urban mix, and in particular includes smaller schools who otherwise would not benefit. However, it also concentrates resources on a core of educationally disadvantaged pupils in urban areas. This results in a further differentiation in resource allocation between pupils in Breaking the Cycle and those in the Scheme of Assistance, while still leaving the majority of disadvantaged pupils without any benefit. This new scheme suggests a policy compromise between two discrete, and conflicting, priorities: a fairer geographic spread of beneficiary pupils and a more intensive targeting of a minority of urban pupils. Similarly, the major expansion of the ABR under the guise of the Local Development Programme, while rigorously targeted, contains 54 per cent of the total population (even more if projects in the non-designated areas are included). This expansion, while logical in terms of the distribution of the unemployed, has the effect of diluting the scale of resources available for intervention in the targeted areas. These attempts to reform targeted poverty programmes serve to confirm the dilemmas inherent in this approach from an efficiency/effectiveness perspective.

ENHANCEMENT OF SERVICE PROVISION

The second rationale for spatial programmes, besides their perceived ability to target those in need, is to provide a focus for enhancing service provision in response to problems of multiple deprivation. Such programmes emphasise local co-ordination and innovative delivery in order to improve the effectiveness of existing resources and services. Their starting point is a recognition that tackling disadvantage requires more than simply re-distributing resources, but must also address the quality of service provision. This is an acknowledgement that welfare services may not be addressing the full extent of people's needs, especially where people who are financially poor are also more likely to be unemployed, have low education levels, experience poor health and live in undesirable housing conditions. It also recognises that involving intended beneficiaries in service planning and delivery is necessary to enhance the effectiveness of services. The Third EU Poverty Programme (Poverty 3) and the Community Development Programme are the two featured examples of area-based welfare programmes which promote inter-agency co-ordination and participative delivery of services. Other examples are the COMTEC pilot programme in the mid-1980s which promoted inter-agency co-ordination and local planning of youth employment services at a county level,[2] and the previously described ABR initiative, which combined area targeting of resources with a focus on local co-ordination of employment services[3] and a recent (1998) pilot initiative for the integrated delivery of state services in disadvantaged urban neighborhoods.[4]

Case Studies: Integrated Service Provision and Community Development

Poverty 3 was an EU initiative between 1989-94, which had the aim of promoting the social and economic integration of socially excluded groups. The programme sought to enhance service co-ordination and delivery based on three organisational principles: multi-dimensionality, partnership, and participation. Two "model action" projects were set-up in Ireland in areas of disadvantage: Forum in north and west Connemara and People Action against Unemployment Limited in Limerick city (PAUL). These localities were chosen on the basis that a high proportion of those living there was in poverty and that they faced other area-specific disadvantages. Thus, PAUL targeted four large local authority estates in the city with the following characteristics:

- A high concentration of unemployed and other low income households, with 86 per cent of tenants dependent on welfare and unemployment rates twice those of the city as a whole (45 per cent as compared to 21 per cent)

- A poor quality of life, with low levels of public and private service provision and also problems of vandalism and political alienation

- Problems of stigma and bad press attached to the estates, creating difficulties for people seeking work and undermining their perceived attractiveness as residential neighbourhoods

- A lack of community structures, together with an exodus of those with leadership abilities (Walsh, 1993; Harvey, 1994).

The problems confronting Forum in Connemara had also a strong geographic dimension, in this instance defined as physical remoteness, small size of farms and poor infrastructure (Harvey, 1994).

Poverty 3 sought to address these issues through targeted interventions which, as well as conferring additional resources to these areas, were integrated with one another and involved local people in their design and implementation. Thus, PAUL established a network of community action centres in its target estates, which were under local control and staffed by people from the area. The centres provided information and advice and gave support to community activity. The project also developed an integrated response to money-lending and indebtedness. In Connemara, Forum piloted a community bus service and a participative health care initiative, and also promoted aquaculture and tourism developments.

The second example of an area-based programme of service enhancement is the Community Development Programme, administered by the Department of Social, Community and Family Affairs. This programme dates from the late 1980s, when it was set up to provide mainstream funding for a number of experimental projects in the Second EU Poverty Programme. The programme is targeted at communities affected by high unemployment, poverty and disadvantage, with the aim of increasing the capacity of local people to improve their situation. (A small number of projects target disadvantaged groups, such as travellers, people with disabilities and women.) The thinking behind the programme is that

> if the marginalisation and powerlessness experienced by people living in urban and rural areas of multiple deprivation is to be tackled, then it is essential to help to make it possible for the people involved to participate effectively and to assist them to act collectively (Dept of Social, Community and Family Affairs, 1998, 15).

The Community Development Programme thus has a specific focus on enhancing the capacity of local people to contribute to and to avail of services, which can be seen to complement the Poverty 3 programme's concern with local inter-agency co-ordination of services.

The programme has expanded greatly since 1990, and now supports almost 90 local projects. These are locally-managed, with widespread voluntary involvement along with two or three paid staff. The projects act as resource centres for a range of community activities and provide personal development opportunities and support services for groups and individuals. Annual core funding is between £40,000 and £60,000. The remit of the projects is not to directly channel additional resources or services to disadvantaged localities, but to enable individuals and groups to articulate their needs and to identify solutions to these needs, in conjunction with other interests. It is intended to eventually extend this programme to all areas of disadvantage and, for this purpose, has used an analysis of Census-derived deprivation indicators to indicate priority locations.

Institutional and Resource Constraints to Local Integration
The targeting dimension of both the Poverty 3 and the Community Development programmes, while apparent in a somewhat ad-hoc fashion, is secondary to its objective of enhancing service provision. This is achieved through the introduction of a new local planning dimension to service provision and the promotion of user empowerment. This area focus is designed to facilitate a shift from traditional sectoral service provision to multi-dimensional strategies that address all aspects of particular issues. A key challenge here is to define an appropriate geographical scale at

such as FAS, the health boards, Enterprise Ireland and the Department of Social, Community and Family Affairs rarely have devolved decision-making capacity. This problem is also acute for business and trade unions, whose involvement in the local partnerships is as representatives of what are primarily national negotiating bodies, with a very weak local dimension. These territorial problems are compounded by the anomalous administrative status of these bodies, in particular their weak connections to local government structures (OECD, 1996).

Finally, it should be noted that these local service initiatives provide little in the way of additional resources or employment in areas of poverty. While they may help people to cope with the poverty better, they can also maintain a cycle of dependency. In this regard, it should be borne in mind that the resource base in disadvantaged areas is very weak, especially in terms of economic resources. As Nolan *et al.* (1998) argue:

> It is crucial that, having contributed to the creation of areas characterised by minimal resources, the state does not develop unrealistic expectations of the extent to which the communities which reside in them can find the solution to the problems which characterise those areas (p. 104).

As such, areas of multiple deprivation require substantial public and private investment and, more importantly, they need to develop mechanisms that retain this added value in the form of sustainable community assets.

PROMOTION OF LOCAL DEVELOPMENT

The third strand of area-based poverty programmes is represented by integrated development initiatives. These programmes have as their starting point a realisation of the deep-rooted and cumulative nature of poverty in certain localities, such that

> exclusion form gainful employment is compounded by other dimensions of social exclusion, particularly the cumulative impact of the concentration of unemployment, poverty and the number of dependants to working population in particular areas and communities. The net effect of such concentration is to isolate communities, families and individu-

which to seek to achieve integration of services (Georis and S monin, 1994). To-date, different scales have been used as the basis for local integration, ranging from individual estates, to neighbourhoods to large urban or rural areas. A complicating factor is that those involved in local projects frequently have diverse territorial remits, covering local community, city, region and nation state. Even at a community level, they may be differences between individual estates and neighbourhoods. Transcending these differences of scale in order to integrate service provision can be problematic, especially for sectoral agencies that have traditionally operated at a wider geographic level. This difficulty is compounded by economies of scale. Thus, while community organisations may wish to have the lowest possible level of service delivery in order to maximise client-focus and access, existing service providers may find that this is impractical on cost grounds. There can also be conceptual differences as to the meaning of integrated service provision (Walsh, 1994b). Integration can mean: (a) local access to services, either through direct local provision or through a one-stop-shop, where a range of service providers can be contacted; (b) local networking of services around a client's needs through a mediator or gatekeeper; or (c) better sharing of information among service providers as to the needs of clients. The new pilot integrated services initiative may provide some clarity as to these issues, given its exclusive focus on this question, though interestingly it omits voluntary and community services (see endnote 4).

An important feature of local service co-ordination programmes is the establishment of new local institutional structures which involve a mix of statutory and community actors (e.g. community development projects, area partnership companies, local employment service networks, model action projects — see Walsh et al., 1997, 1998 for a description of these). The modus operandi of these structures has been subject to considerable scrutiny, which will not be repeated here (Kelleher and Whelan, 1992; Cullen, 1994; Curtin and Varley, 1995; OECD, 1996; Walsh et al., 1997, 1998). There is, however, an important spatial dimension which is worth highlighting: the degree to which agencies have the autonomy to devise and plan services at the local level, even where the have the desire. Local personnel in regional and national bodi als from the mechanisms which sustain social cohesion. The consequences are manifested in environmental decay, demoralisation, social breakdown and crime as well as high concentrations of unemployment (Ireland, 1995, 30).

Disadvantaged areas are therefore seen as being linked into a cycle of economic and social decay that can only be reversed if the indigenous development potential of these areas is properly harnessed. Central to this approach is an enhancement of the capacity of disadvantaged communities to become the prime movers in local economic development. This takes the emphasis in previous area programmes on community development a stage further by linking it explicitly with local economic self-help and enterprise. This approach is referred to as bottom-up or integrated (socio-economic) development and the main examples are:[5]

- The Global Grant for Local Development (an initiative linked to the ABR programme discussed above)
- The Integrated Development in Designated Areas of Disadvantage Programme (IDDADP) (part of the Local Development Programme)
- The EU Urban Programme
- The EU Peace and Reconciliation Programme for Northern Ireland and the southern border counties.

These initiatives incorporate a stronger emphasis on economic activities than previous poverty programmes. The influence of the EU in formulating these programmes is twofold: first, by providing the additional resources to support these initiatives, in particular the use of structural funds, and second, by supporting the emphasis on local economic and enterprise development. The focus here is on the IDDADP and the Urban programme.

Case Studies: Local Development and Urban Regeneration

The IDDADP runs from 1995 to 1999. The objective of the programme is

> to counter disadvantage through support for communities which make a collective effort to maximise the development

potential of their areas, which are capable of a sustained effort to implement a plan and which have committed an appropriate level of resources, broadly defined, to that process (Ireland, 1995, 59).

IDDADP has two broad measures: enterprise and infrastructure (representing economic capital); and education and training, services for the unemployed and community development (reflecting the importance of social capital). Area-targeting is incorporated through the prior government designation of 33 disadvantaged localities across the country based on Census-derived indicators of need and administrative factors (Haase in this volume outlines how these indicators were developed and applied). Subsequently, 38 local development partnerships were established in the designated areas to deliver the programme on the basis of a local development plan. These included the original 12 projects under the ABR initiative and others supported under the Global Grant for Local Development. A minor amount of programme funding is allocated to 33 community projects in non-designated areas. This programme of socio-economic development is managed by an intermediary company, Area Development Management, which is responsible for animating local structures, funding action plans, monitoring outcomes and identifying policy and practice issues. The total funding is around £100m, with approximately £2m per partnership over a four-year period (and lesser amounts for community groups). Again, it is intended that these funds are supplemented by resources drawn from local agencies, including from private sector interests. The main achievements of the programme after two years (end 1997) are:

- 4,400 micro-enterprises employing 7,800 previously long-term unemployed people
- 9,500 people placed in employment
- 16,000 disadvantaged children on preventive educational projects
- 8,600 unemployed adults participating in education and training programmes
- 200 amenity and environmental projects

- 500 groups and communities benefiting from community development (Area Development Management, 1998).

A second example is the EU Urban Community Initiative Programme (1996-99). Urban is directed at large cities (100,000+) containing neighbourhoods which suffer from lack of economic opportunity, low incomes, a generally poor quality of life and a deteriorating urban fabric. These disadvantaged neighbourhoods, which can be identified geographically by various socio-economic indicators, are seen as arising from

> a complex network of inter-related factors including high unemployment and social welfare dependency, mass housing development in peripheral areas, decaying urban infrastructures, low education and skill levels and lack of economic opportunities (Dept of the Taoiseach, 1996, 6).

In response, Urban proposes an integrated approach that includes supporting small businesses, enhancing local infrastructure, improving the management of public housing, providing customised training, developing social amenities and promoting equal opportunities. The long-term aims are to bring about a lasting improvement in the living standards of inhabitants and to develop more effective systems of urban governance.

In Ireland, three urban areas have been selected for inclusion in Urban: the northside of Cork city, north Dublin city (Finglas, Ballymun and Darndale) and Tallaght/Clondalkin in south county Dublin. The total funding allocated under this initiative in Ireland is £20m, with half of this going to north Dublin. New institutional structures involving a range of interests have been set up in each area to administer the programme, with a lead role for the local authority. Urban is still at an early stage of implementation, with considerable delays associated with its infrastructural projects.

Sustaining Economic Development in Resource-poor Areas

How has the spatial dimension been operationalised in these programmes to promote local socio-economic development in areas with high levels of poverty? Again, the starting point is the designation of certain areas as disadvantaged. This is most extensive under the IDDADP, with 54 per cent of the total population being included (1.9m people), not including those included in the non-designated areas. An interesting feature in this designation was the decision to include whole counties in rural areas, though in some cases only containing relatively small areas of disadvantage. The 33 additional localities selected for lesser funding under this programme were not designated in advance, but reflected local responses to disadvantage. Within these broad parameters, local development partnerships are expected to further target their resources at specific sub-areas and groups, though how this is operationalised is left to local discretion. By contrast, the Urban programme is confined to three sub-areas in Dublin and Cork (ranging in population between 55,000 and 115,000), while an estimated 30-40 localities in urban areas with a population of over 4,000 are expected to be included under the Urban Renewal Scheme. In both these instances, areas were put forward for selection at a local level using broad national criteria. Other common aspects of these local development programmes as compared to early area programmes are their focus on integrated local planning and the establishment of new local institutions (Indeed, these features of the precursor ABR and Poverty 3 initiatives were simply incorporated into the IDDADP.)

The innovative feature of local development programmes is their emphasis on generating local economic development in disadvantaged areas. The main instruments in this regard are support for micro-enterprises and training and employment opportunities. This leaves an imbalance between investment in human capital and funding for physical and economic assets, e.g. housing, amenities, environment and transport. Both Urban and, to a lesser degree, the IDDADP, have an emphasis on the effective management of physical resources, e.g. public housing. However, the construction of social housing remains exclusively within the ambit of the Department of the Environment and Local Government and the local authorities. There is some expectation that the new Urban Renewal Scheme may address this gap between social development and physical renewal. However, it remains to be seen whether it will be possible to extract community gain from what is an exclusively private investment scheme, with no direct public funding. There is a danger that critical elements of the physical infrastructure in disadvantaged areas, such as housing and amenities, will remain neglected under this scheme.

A related weakness in these local development initiatives is the absence of community structures that can mobilise resources for economic and physical development. Local development partnerships are not used as vehicles for major public or private investment in local infrastructure. This contrasts with the UK, where similar structures act as conduits for public-private construction projects, while community development trusts are widely used as a means of mobilising investment in community assets. There is also a dearth of alternative financial instruments that could underpin economic development in disadvantaged areas. In other countries, community credit unions, community banks and social investment trusts provide a means for enhancing the financial resources of disadvantaged urban areas. However, the establishment of similar finance bodies has not been a feature of Irish local development programmes in disadvantaged areas, although individual credit unions have contributed in this regard. Yet there is considerable evidence of a widespread "financial desertification" of disadvantaged areas, with a knock-on impact on the long-term economic development of these localities (Leyshon and Thrift, 1994). Interestingly, a number of integrated area plans under the 1998 Urban Renewal Scheme suggests the application of a levy on commercial development as a community investment fund.

CONCLUSION: ENHANCING THE SPATIAL DIMENSION IN AREA-BASED ANTI-POVERTY PROGRAMMES

This review has charted the growing popularity of spatial anti-poverty programmes in welfare policy. These programmes incorporate varying spatial aspects, with a span from a simple area targeting to local co-ordination of services to integrated local development. This review identified a number of fundamental weaknesses in these approaches. As recent research (reported by Haase and Williams *et al.* in this book) shows, unemployment and poverty are spatially pervasive phenomena. While many areas have above average high rates of unemployment, the incidence of unemployment remains spatially quite diffuse. In order to get to a stage where a majority of the unemployed is covered through using DEDs, the total population base has to be increased to 40 per cent. To achieve a higher proportion of the unemployed means an ever further dilution of the targeting principle. As well as limitations on the share of the poor population that benefits from targeting, there are also concerns about the quality and effectiveness of simply providing additional resources. There may also be negative implications for those who do not benefit in terms of actually worsening their situation as area targeting may simply displace social problems from one location to another. Such displacement has been a very visible outcome of the urban renewal scheme, where new zones of urban blight have emerged on the fringes of designated areas (KPMG et al., 1996). To what extent this has also occurred in social programmes is harder to say, given the lack of comparable information. Much depends here on the overall level of opportunity and whether this has been expanded or not. For example, in an era of growing employment, improving job opportunities in one area may not reduce opportunities for the unemployed in other areas.

Other, more sophisticated, forms of local intervention have greater potential in terms of addressing poverty. However, even here, the review has highlighted some drawbacks. Thus, programmes of local service co-ordination are limited by their *ad hoc* institutional status in an otherwise centralised system of public administration, and also by lacking any linkage with a rather weak local government structure. There are also technical and cost limitations to achieving local co-ordination of services, while community empowerment without a parallel shift in the operation of state agencies will have limited impact. Meanwhile, local development initiatives that do not address the weak economic resource base in communities are unlikely to succeed in the longer term.

At a wider level, this highlights the importance of locating local problems within a national policy framework that addresses structural deficiencies in how society allocates its resources (e.g. polarisation in housing, pupil selectivity between schools, differential access to public and private services and inadequate provision of public transportation). This highlights the importance of the recently-announced National Anti-Poverty Strategy, which seeks to make poverty a core concern of all aspects and levels of public policy making. Only when there is an enhanced national framework for tackling poverty will the potential of local interventions be maximised.

It is also clear then that the spatial dimension in welfare programmes requires further elaboration. In general, location is still treated as a residual element, whether as a cost-efficient way to ring fence disadvantaged groups, as a milieu for local co-ordination of and community involvement in service provision, or as a rallying point for promoting "bottom-up" development. Where then does this leave the spatial dimension in anti-poverty programmes as part of overall welfare policy? One avenue would appear to rest on identifying and addressing the cumulative effect that may occur in localities with high concentrations of poverty. For example, Haase et al. (1996) argue for the existence of a distinct spatial dimension to economic and social forces which

> is increasingly seen to have an independent effect on the degree of social exclusion experienced by those living in (disadvantaged) areas. This "cumulative effect" has been clearly demonstrated in, for example, the area of educational research, where is has recently been shown that the scholastic achievement of students from equally deprived family backgrounds differs markedly depending on whether they attend an average class/school or share the school environment with students from equally deprived family backgrounds (p. 9).

Similarly, NESC has identified a form of cumulative localised disadvantage:

> It is this localised process of interaction between labour market, education, housing and environmental factors which is most likely to be addressed effectively by an area-based strategy. . . area-based strategies can complement national policies designed to stimulate employment growth and thus address mechanisms which exclude particular communities or populations from benefiting from generalised improvements in economic and employment conditions (NESC, 1993, 417).

Despite these arguments, the lack of research on the additional exclusionary effects in high-poor localities has stymied attempts to strengthen the spatial dimension in welfare policy. Current research by the Combat Poverty Agency and the Howard Foundation in examining living conditions in local housing estates, in particular, those factors which precipitate a cycle of decline in so-called "dump" estates, may point a way forward. Also, the ESRI is doing further analysis based on the Living in Ireland Survey on the cumulative effects of location. These are only a start: there remains an inadequate understanding of the locational aspects of disadvantaged areas. The further involvement of geographers in researching the substance of area-based policies would assist here.

It is also important that there should be a stronger cognisance of the spatial dimension in public policy. Currently, there exists an uneasy dichotomy between national policies on one hand and selective local interventions on the other. It should be recognised that all public policy has an inherent spatial dimension, whether it is through agricultural policies which benefit commercial producers, industrial policies which discriminate in favour of larger urban centres, housing policies which enforce a pattern of public-private segregation or the policies of financial institutions which deny access to financial services in certain locations. This neglect is made worse by the lack of an explicit urban, rural or regional focus to government policy. These spatial categories provide an intermediate policy zone between the local and the national, which would offer considerable potential in terms of addressing the negative outcomes of socio-spatial processes in specific localities. The next round of the structural funds presents an important opportunity to strengthen the spatial dimension in public policy. However, this should go far beyond some form of artificial regionalisation in order to target resources. Rather, it involves the formulation of comprehensive policies to promote balanced, sustainable and socially inclusive development at rural, urban and regional levels.

NOTES

1. Information on the number of beneficiaries and the cost of the schemes is sourced from the response to a Dáil question by the Minister for Education and Science, 25 November 1997.

2. The COMTEC (Community Training and Employment Consortia) programme was initiated in 1985 by the Youth Employment Agency (now part of FAS), as a local co-ordination and planning mechanism for youth employment services. Eight COMTECs were established at the equivalent of a county level to formulate and implement a two-year plan. The COMTECs consisted of a consultative council and a planning unit representing the main service providers. The programme was designed to facilitate three processes: coordination between agencies, local participation in the planning of services, and implementation of a local plan. Minimal resources were provided for the initiative (£350,000 or £20,000 per project per year). The initiative was wound up in 1987 (Joyce and Daly, 1987).

3. The ABR emphasised integrated activity, defined in this instance as (a) linking training, education and work experience services and (b) relating these to opportunities for employment and enterprise. This included both inter-agency co-ordination and promoting new services. The starting point for this was an "initial contact" programme whereby the ABR projects assessed the needs of the long-term unemployed and directed them to appropriate training. The model of intervention has since been mainstreamed as the Local Employment Service (Task Force on Long-term Unemployment, 1995).

4. The aim of this initiative is to develop new procedures for the co-ordination of statutory service delivery, which can then be mainstreamed as models of best practice. Four neighbourhoods have been selected for this pilot phase, based on ranking of disadvantaged urban areas using Census-derived indicators of disadvantage. The first phase of the programme is to profile existing services, to assess the extent of integration and to identify duplication and gaps. This information will then be used to develop an action plan to enhance local service co-ordination. There also exist two other service co-ordination initiatives dealing with access to labour market programmes/job placement/counselling and guidance (Local Employment Service) and indebtedness/access to credit/money management/consumer rights (Money Advice and Budgeting Service).

5. Also of relevance in terms of local development are the County Enterprise Boards, the Leader programme and the 1998 Urban Renewal Scheme. While neither have an explicit disadvantage focus, there is considerable overlap with local development initiatives in disadvantaged areas. For example, the Urban Renewal Scheme encourages private sector investment in the physical development in rundown urban areas through the granting of tax reliefs. Following a review of the scheme in 1996 (KPMG *et al*, 1996), the current

scheme has adopted an integrated approach to urban regeneration, covering social, physical, economic and environmental issues in designated areas (Department of the Environment and Local Government, 1997). Central to this new approach is the preparation of integrated area plans under the leadership of local authorities. As well as targeting urban areas with high levels of disadvantage, the new urban renewal scheme is expected to deliver tangible benefits for disadvantaged communities in terms of employment opportunities, community facilities and social housing.

REFERENCES

Area Development Management (1998) *Reaching Out To The Excluded, A Summary Report On Progress In 1997*, Area Development Management, Dublin.

Callan, T., Nolan, B., Whelan, C. and Williams, J. (1996) *Poverty in the 1990s*, Oak Tree Press, Dublin.

Combat Poverty Agency (1989) *Towards a Funding Policy for Community Development*, Combat Poverty Agency, Dublin.

Combat Poverty Agency and Community Workers Co-operative (1990) *Community Work In Ireland: Trends In The 80s, Options For The 90s*, Combat Poverty Agency and Community Workers Co-operative, Dublin.

Combat Poverty Agency (1993) Unpublished working papers on identifying prospective areas for inclusion in the local development programme.

Commission of the European Communities (1993) *Growth, Competitiveness and Employment; The Challenges and Way Forward into the 21st Century*. Office for Official Publications of the European Communities, Luxembourg.

Commission of the European Communities (1995) *A European Strategy for Encouraging Local Development and Employment Initiatives*, COM (95), 273.

Craig, S. and McKeown, K. (1994) *Progress Through Partnership, Final Evaluation Report on the PESP Pilot Initiative on Long-Term Unemployment*, Combat Poverty Agency, Dublin.

Cullen, B, (1989) *Poverty, Community and Development*, Combat Poverty Agency, Dublin.

Cullen, B, (1994) *A Programme in the Making, A Review of the Community Development Programme*, Combat Poverty Agency, Dublin.

Curtin, C. and Varley, T. (1995) Community action and the state, in Clancy, P., Drudy, S., Lynch, K. and O'Dowd, L. (eds.) *Irish society: sociological perspectives*, Institute of Public Administration and the Sociological Association of Ireland, Dublin.

Department of the Environment (1991) *A Plan for Social Housing*, Department of the Environment, Dublin.

Department of the Environment and Local Government (1997) *1998 Urban Renewal Scheme — Guidelines*, Department of the Environment and Local Government, Dublin.

Department of Social Welfare (1997) *Supporting Voluntary Activity, A Green Paper on the community and Voluntary Sector and Its Relationship With the State*, Stationery Office, Dublin.

Department of Social, Community and Family Affairs (1998) *Supports for Community Development and Family Support Groups, 1988-1999*, Department of Social, Community and Family Affairs, Dublin.

Department of the Taoiseach (1995) Operational Programme Urban — Ireland, 1996-1999, unpublished report.

ESRI, DKM Consultants, Boyle, G. and Brendan Kearney and Associates (1993) *EU Structural Funds, the Community Support Framework: Evaluation and Recommendations*, Stationery Office, Dublin.

Georis, P. and Simonin, B. (1994) The multi-dimensional approach as a principle of the programme, in EEIG Animation and Research, *The Lessons of the Poverty 3 Programme*, EEIG Animation and Research, Lille.

Haase, T., McKeown, K. and Rourke, S. (1996) *Local Development Strategies for Disadvantaged Areas, Evaluation of the Global Grant, 1992-1995*, Area Development Management, Dublin.

Harvey, B. (1994) *Combating Exclusion: Lessons from the Third EU Poverty Programme in Ireland, 1989-94*, Poverty 3 Programme in Ireland, Dublin.

Interdepartmental Group on Urban Crime and Disorder (1992) *Report*, Stationery Office, Dublin.

Ireland (1995) *Operational Programme for Local Urban and Rural Development, 1994-1999*, Stationery Office, Dublin.

Ireland (1997) *Sharing in Progress, The National Anti-Poverty Strategy*, Stationery Office, Dublin.

Joyce, L. and Daly, M. (1987) *Towards Local Planning: An Evaluation of the Pilot COMTEC Programme*, Institute of Public Administration, Dublin.

Kelleghan, T., Weir, S., O hUalachain, S. and Morgan, M. (1995) *Educational Disadvantage in Ireland*, Department of Education, Combat Poverty Agency and Educational Research Centre, Dublin.

Kelleher, P and Whelan, M. (1992) *Dublin Communities in Action: A Study of Six Projects*, Combat Poverty Agency and Community Action Network, Dublin.

KPMG, Murray O'Laoire, NIERC (1996) *Study on Urban Renewal Schemes*, Stationary Office, Dublin.

Leyshon, A and Thrift, N. (1994) Access to financial services and financial services withdrawal: problems and policies, *Area*, 26(3), 268-275.

McCashin, T. (1990) Local communities and social policy, in Combat Poverty Agency and Community Workers Co-operative (1990) *Community Work in Ireland: Trends in the 80s, Options for the 90s*, Combat Poverty Agency and Community Workers Co-operative, Dublin.

Ministerial Task Force on Measures to Reduce the Demand for Drugs (1996) *First Report*, Stationery Office, Dublin.

NESC (1990) *A Strategy for the Nineties: Economic Stability and Structural Change*, National Social and Economic Council, Dublin.

NESC (1993) *A Strategy for Competitiveness, Growth and Employment*, National Social and Economic Council, Dublin.

NESC (1994) *New Approaches to Rural Development*, National Social and Economic Council, Dublin.

NESC (1995) *Quality Delivery of Social Services*, National Economic and Social Forum, Dublin.

Nolan, B., Whelan, C. and Williams, J. (1994) Spatial aspects of poverty and disadvantage, in Nolan, B. and Callan, T. (eds.) *Poverty And Policy*, Gill and Macmillan, Dublin.

Nolan, B., Whelan, C. and Williams, J. (1998) *Where Are Poor Households? The Spatial Distribution of Poverty and Deprivation in Ireland*, Oak Tree Press in association with the Combat Poverty Agency, Dublin.

O Cinneide, S. (1985) Community response to unemployment, *Administration*, 33(2), 231-257.

O Cinneide, S. and Walsh, J. (1992) Multiplication and divisions: trends in community development in Ireland since the 1960s, *Community Development Journal*, 25(4), 326-336.

Organisation for Economic Co-operation and Development (1996) *Ireland — Local Partnerships and Social Innovation*, OECD, Paris.

Powell, F. (1992) *The Politics of Irish Social Policy, 1600-1990*, Edwin Mellen Press, Landpeter.

SUS Research (1987) *CODANS (County Dublin Areas Of Need)*, Dublin County Council, Dublin.

Task Force on Long-term Unemployment (1995) *Interim Report*, Stationery Office, Dublin.

Walsh, J. (1993) Pioneering a strategy for integrated urban development — the PAUL Partnership, Limerick, in Commins, P. (ed.), *Combating Exclusion in Ireland, 1990-94*, Poverty 3 Programme in Ireland, Dublin.

Walsh, J. (1994a) *Report on the Implementation of the Poverty 3 Programme by the PAUL Partnership Limerick*, PAUL Partnership Limerick, Limerick.

Walsh, J. (1994b) Multi-dimensionality — theory into practice, *Poverty Today*, 26.

Walsh, J., Craig, S. and McCafferty, D. (1997) *The Role of Local Partnerships in Promoting Social Inclusion — Ireland*, European Foundation for the Improvement in Living and Working Conditions, Dublin and Office for Official Publications of the European Communities, Luxembourg.

Walsh, J., Craig, S. and McCafferty, D. (1998) *Local Partnerships and Social Inclusion*, Oak Tree Press in association with the Combat Poverty Agency, Dublin.

Part 5

Conclusion

14

POOR PEOPLE, POOR PLACES: CONCLUSION

Dennis G. Pringle
National University of Ireland, Maynooth

Jim Walsh
Combat Poverty Agency, Dublin

Poverty clearly has a "geography": some areas contain a higher percentage of households experiencing poverty than other areas, whilst the ways in which poor households experience poverty also varies from place to place. However, poverty, when considered at all by academics, is generally analysed from a sociological, economic or anthropological perspective: it rarely provides the central focus for geographical enquiry. Nevertheless, geographers have collectively developed a reasonable understanding of the geography of poverty in Ireland through research primarily focused on other topics in which social inequality tends to be a recurring motif (e.g. accessibility to services, inequalities in health, local and regional development issues, housing problems, etc.). Much of this information remains diffuse and scattered; much of it also remains contested. As a result, the geographical dimension of poverty tends either to be ignored in public policy or else incorporated in a residual manner, although the National Anti-Poverty Strategy (1997) provides a notable recent exception in this regard. The objective of the *Poor People, Poor Places* conference in Maynooth was to collate some of this information and to initiate a debate, involving both geographers and decision makers in the public sector, on the nature and causes of the geography of poverty in Ireland, and the policy implications for intervention.

The preceding chapters in this book provide a variety of observations and analyses of poverty from a geographical perspective. This chapter identifies some of the major issues which have been raised.

POVERTY, DEPRIVATION AND SOCIAL EXCLUSION

The terms "poverty" and "deprivation" are sometimes used in everyday language as inter-changeable synonyms for one another, whilst the expression "social exclusion" is sometimes loosely used as a synonym for both. Most authors in the preceding chapters have stressed the need to make a clear distinction between these concepts and have defined their own understanding of the terms which they use. Although some differences in interpretation may be observed, there would appear to be a high degree of consensus between most authors. The core issue here is how one defines and measures poverty, deprivation and social exclusion in an affluent country in a way that is valid, meaningful and applicable to policy (Nolan and Whelan, 1996). This section attempts to identify a formal conceptual framework, based upon some of the points made in the previous chapters, to facilitate later discussions.

Poverty is regarded by most authors as an income-related concept — i.e. people in poverty have lower amounts of disposable income (or spending power) than people not in poverty. The extent to which a poor person falls below the acceptable standard can be expressed (at least in theory) in monetary terms. *Deprivation*, in contrast, is a more diffuse concept related to the quality of life — i.e. deprived people have a reduced access to various features which other people regard as "normal", if not "essential", for a reasonable quality of life. Deprivation is multidimensional given there are so many different things which people may be deprived of. This multidimensionality makes deprivation difficult to quantify. For example, if attempting to quantify deprivation, how much weight should one give to the absence of an indoor toilet compared with poor accessibility to a maternity clinic? The problem of deciding upon an appropriate weighting is compounded by the fact that the relative importance of these features is likely to vary considerably from one person to another (e.g. an elderly bachelor compared with an expectant mother). Unlike poverty — the extent of which can be quantified in monetary units — there are no obvious units for quantifying the extent of deprivation.

Poverty is defined relative to a norm — i.e. people are classified as being in poverty if their disposable income falls below a specified cut-off value. This cut-off value may be defined either in *absolute* terms (i.e. the income required to meet the requirements for a defined standard of living) or in *relative* terms (e.g. relative to the national average). Most authors in this volume favour a relative conception of poverty, mainly due to the fact that it is difficult to identify a meaningful absolute norm. Very few people in Ireland — unlike those in some Third World countries — do not have access to the essentials to sustain life (i.e. food, clothing and shelter), therefore what people in Ireland would regard as "essential" for a "reasonable" standard of living is much greater than that by people living in some of the poorer Third World countries. "Absolute" definitions of what is essential, when examined more closely, are not really "absolute" but are "relative", in the sense that they are socially determined and vary between cultures; also, within any one culture, they vary over time — i.e. we have a much higher evaluation of what is essential for a reasonable standard of living today than did our parents or grandparents.

A relative conception of poverty avoids the need to make subjective decisions about what is or is not essential for a defined standard of living. However, it requires one to make other subjective decisions. For example, if the poverty line is to be defined as a percentage of the mean national income, then what percentage should be used as the cut-off value: fifty per cent, sixty per cent, one hundred per cent? Should income be measured on an individual or household basis? How should one adjust for variations in household size, and the economies of scale which reduce per capita costs in larger households? Should income measures take account of non-discretionary deductions (e.g. taxation) and benefits in kind? There are many thorny issues associated with the operationalisation of relative concepts of poverty, but the concept itself is clear enough.

Although perhaps obvious, it should be noted that relative poverty is more a function of *social inequality* than it is of social wealth — i.e. it is a reflection of the distribution of wealth within society; not the total, or even the average, wealth of a society. For example, if the income of the wealthier income groups increases at a faster rate that that of the lower income groups during a period of prosperity, then there may well be an increase in the number of households living in relative poverty, even though the poor may be better off in absolute terms than they were previously. Likewise, the percentage of households in relative poverty could decline during an economic recession if the economic burdens of the recession were borne disproportionately by the more affluent sections of society. Some would argue that some increase in relative poverty would be an acceptable price to pay if it resulted in an increase in absolute income for everyone; others would argue that any increase in social inequality in unacceptable, not only on moral grounds, but also because of the documented negative implications of increasing social inequalities upon crime rates, suicide rates and health (e.g. McCloone and Boddy, 1994; Wilkinson, 1996). The debate between those favouring an emphasis upon economic growth and those favouring greater social equality lies at the heart of the right-wing/left-wing cleavage which dominates the party political system in most west European states (Rokkan, 1970; Taylor and Johnson, 1979; 1993).

Deprivation can be measured either using *objective* criteria (e.g. the presence or absence of various features regarded as essential to an acceptable quality of life) or people's *subjective* self-assessments of whether they feel deprived or not. Objective indicators provide a better gauge of "who has what", but subjective measures may provide a better indication of people's satisfaction with what they have. Both types of measure have an important role to play in the development of our understanding of deprivation. There is generally a reasonable degree of correspondence between objective and subjective measures. For example, Storey in this volume noted that 'It can be concluded that there is a considerable degree of correspondence between so-called "objective" assessments of living conditions and the subjective views of respondents, with the majority of those deemed deprived according to the "objective" analysis regarding themselves as poor.' However, there are also some significant differences. For example, McCafferty and Storey each reported case studies in which people classed as deprived by objective indicators did not perceive themselves to be deprived. This may indicate limitations in the ability of objective indicators to correctly identify need, but it could also indicate that people who are most in need may have lower expectations and are therefore more likely to feel content with what they have than others who are objectively better off. Subjective indicators are often informative, but objective indicators probably provide a more reliable indication of social inequalities.

Objective deprivation, as previously noted, can take many different forms. It is useful to make a distinction between *material* and *non-material* forms of deprivation, where material deprivation refers to the absence of material goods (e.g. television sets, cars, washing machines) regarded as essential for an acceptable quality of life, and non-material deprivation refers to the absence of any other feature which impacts upon the quality of life. Non-material forms of deprivation include physical illnesses and disabilities, impaired social networks (resulting, for example, in loneliness), and psychological and emotional problems (e.g. low self-esteem, fear of attack). Non-material forms of deprivation are as varied as material forms of deprivation, but they are less tangible and more difficult to quantify. They consequently tend to receive much less attention than material deprivation. Indeed, we appear to live in a society obsessed by economics and material possessions, which induces a form of tunnel vision which causes us to overlook many of the more important influences upon human happiness. One of the challenges for geographers and other social scientists is to develop more powerful tools for capturing the rich tapestry of life associated with specific places.

Several authors make an important distinction between *actual* and *potential* deprivation, where the term "potential deprivation" is used to refer to groups (such as the elderly and single parents) who are more likely to experience actual deprivation. The per capita rate of actual deprivation amongst these groups is higher than it is amongst the general population, but it does not follow that all members of these groups are deprived — many elderly people, for example, are quite affluent (and healthy), so being old *per se* is not a form of deprivation. There is some disagreement about whether composite deprivation indices should or should not include measures of potential deprivation. Some researchers (e.g. SAHRU, 1997) argue that deprivation indices should be based solely on measures of actual deprivation, whereas others (such as

Haase in this volume) argue that it is valid to include measures of potential deprivation on the grounds that high rates of potential deprivation provide an indirect indication of high rates of actual deprivation. The validity of these arguments will depend upon the extent to which members of a sub-group experience a higher rate of actual deprivation, the degree of spatial concentration of these sub-groups, and whether the risks of actual deprivation are spatially invariant (e.g. if elderly people living in affluent areas have a low risk of actual deprivation, then using the percentage of people who are elderly as a measure of potential deprivation could be misleading). More research is required to shed light on the nature of the relationships between potential and actual deprivation.

It is not always easy to make a clear distinction in practice between potential and actual deprivation. For example, should unemployment be regarded as a form of potential or actual deprivation? Unemployment is generally associated with low disposable incomes, but some unemployed people (especially those who have received generous redundancy settlements, or who live in households with other income sources) may be relatively affluent, so (viewed in narrow economic terms) it could be argued that unemployment should be regarded as an indicator of potential rather than actual deprivation. On the other hand, if one adopts a broader concept of deprivation to include psycho-social aspects such as demoralisation, then it could be argued that unemployment, especially long-term unemployment, should be regarded a form of actual deprivation.

The relationship between poverty and deprivation can be viewed in different ways. Cook *et al.* suggest that poverty is a cause of deprivation — i.e. it is impaired spending power which forces people to forgo some of the features which are regarded by others as an essential component of the quality of life. Poverty is a form of resource constraint, whereas deprivation refers to the implications of this resource constraint upon consumption. Others tend to regard poverty as a function of deprivation. For example, unemployment (a form of deprivation) increases the likelihood of poverty. It is suggested here that this apparent contradiction with regard to cause and effect can be resolved by making a distinction between potential and actual deprivation (Figure 14.1). Poverty may be regarded as the intervening link between potential and actual deprivation. Groups, such as single parents and the elderly, run a greater risk of poverty (whether defined in absolute or relative terms); poverty, in turn, increases the likelihood of actual material deprivation. Poverty also increases the likelihood of non-material deprivation (such as poor health, or demoralisation), but other factors (e.g. age, declining population due to out-migration) are likely to exert a more direct influence upon non-material forms of deprivation. However, these other factors are themselves likely to be more prevalent amongst the same groups of people as those most likely to experience poverty.

Figure 14.1: The Poverty/Deprivation Nexus

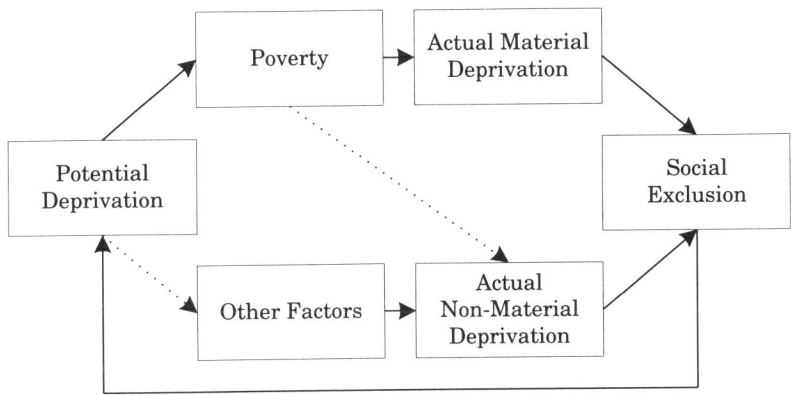

It will be noted that the model also includes links between both material deprivation and non-material deprivation and *social exclusion*. Social exclusion is defined differently by different authors. At least three different meanings may be identified. Its most basic application is as a vogue term to refer to deprivation/poverty, especially to emphasise its multi-dimensional nature. However, it is used in this chapter as including anything which inhibits full participation in society.

Social exclusion, in its original French derivation, has a relational focus: it refers to social participation, social integration and access to power (Room, 1995). This second meaning emphasises social exclusion as a dynamic force, which is based on social processes and relationships. If we take education as an example, low educational attainment is not only a form of material deprivation, it also reflects a dynamic within the educational system whereby a minority of pupils are systematically marginalised and excluded through practices of school segregation, streaming, emphasis on educational performance, and hidden educational costs.

A third meaning of social exclusion is its emphasis on the collective and spatial dimensions of deprivation. This refers to the way in which social relationships are influenced by spatial relationships, through for example the planning system or the media. These socio-spatial relationships can be perceived through local case studies, such as those reported in this volume by Bartley and McCafferty. This perspective highlights the social management of deprived areas, ranging from the provision of financial services, to control of social order, to media portrayals (for example, references to an urban "underclass"), as being as added dimension to the concentrated levels of deprivation evident in certain areas (Madanipour *et al.*, 1998). It also emphasises the importance of social capital in an area, defined by Puttnam (1995, 664) as the "networks, norms and trust that enable participants to act together more effectively to pursue shared objectives". This is not to infer some type of gemeinschaft/gesellschaft split, with a narrow policy emphasis on community re-invigoration. Rather it is to reassert the significance of the community as a local agency in mediating the processes of social exclusion.

Social exclusion may be regarded as a form of deprivation (in the sense that people who are socially excluded are deprived of the things from which they are excluded), but social exclusion goes beyond that. Whilst poverty and material deprivation may be regarded as indicators of social inequality, social exclusion is related more to the concept of power (or, to be more precise, the lack of power). The socially excluded are not only deprived of material or non-material features regarded by others as essential to a reasonable quality of life; they are also deprived of a significant influence in controlling the processes that give rise to these inequalities. Whether having a significant input would be sufficient to make much of an impact upon the structural inequalities discussed by MacLaran and others in this volume is a moot point, but the absence of a significant input does little to reduce the vulnerability of the groups most at risk. This is represented in Figure

14.1 by the feedback loop from "social exclusion" to "potential deprivation".

The model depicted in Figure 14.1 does not attempt to identify every possible causal path — only the major hypothesised causal links are identified. For example, actual material deprivation will obviously be influenced by factors other than poverty, but the inclusion of a link between "other factors" and "actual material deprivation" (and other similar links) would only tend to distract from what are believed to be the more significant features of the model.

MEASURING AND MAPPING POVERTY AND DEPRIVATION

The first step in many geographical analyses is to map the spatial distribution of the phenomenon being studied. This, in turn, requires accurate measurements of the phenomenon for various locations. The measurement and mapping of poverty and deprivation raises a number of issues.

In most instances we would ideally prefer to measure poverty, as poverty is the major determinant of material deprivation. However, comprehensive income data are rarely available. There are some exceptions, most notably the 1987 Survey of Income Distribution, Poverty and Usage of State Services and the Living in Ireland Surveys conducted by the Economic and Social Research Institute since 1994. Findings from these surveys are reported by Nolan et al. (1998) and, in this volume, by Williams et al. However, surveys of this type are expensive and extremely time-consuming. Also, despite the relatively large size of the samples taken (3,294 and 4,048 households in 1987 and 1994 respectively), and the extensive range of information gathered, these surveys provide only limited information on the spatial distribution of poverty and wealth — samples of this size are equivalent to approximately only one household per DED. Also the extent to which these samples may be regarded as representative, irrespective of the many strenuous safeguards implemented by the survey organisers, must be questioned given that they had 36 and 37.5 per cent non-response rates (which, incidentally, are by no means untypical for surveys of this nature — Kawachi and Kennedy, 1997). These surveys provide the best information currently available about the distribution of income and wealth in Ireland, but if the government is serious about tackling poverty and social inequality in Ireland then maybe the time has come to investigate the possibility of collecting more comprehensive sources of information. One possible strategy (although by no means non-problematical) might be to consider adding an income question to the quintennial Census of Population, as in the United States, Canada, Israel and other countries. Another option might be to utilise other sources of information (e.g. tax returns, welfare dependency records, etc.) to produce aggregate data (e.g. estimates of mean income, or percentage of households in relative poverty) for DEDs or other small areas.

In the absence of comprehensive sources of information on income, it is necessary to fall back upon the use of deprivation indicators as surrogate measures of poverty. One advantage in using deprivation indicators is that comprehensive sources of data are available in the form of the Small Area Population Statistics (SAPS) derived from the Census of Population by the Central Statistics Office. However, the SAPS data contain only a limited range of suitable indicators of actual deprivation, so most studies consequently supplement the indicators of actual deprivation with indicators of potential deprivation (such as the percentage of elderly persons, social class measures, housing tenure measures, and so forth). Although many of these features are spatially correlated with poverty and actual deprivation, their inclusion in a composite deprivation index tends to cloud the distinctions which should be made between the "dependent" and "independent" variables. For example, consider what might be an appropriate policy response if a composite deprivation index identified an area as deprived because it contains a large percentage of old people. One obvious response would be to introduce services which alleviate the problems of older people in the deprived area. This would hopefully improve the quality of life for the old people living in the deprived area, and therefore reduce the level of actual deprivation. Now consider what impact such policies would have upon the composite deprivation index. The answer is it would probably have no effect whatsoever because the elderly people would still exist in the same numbers as before and, although they would no longer experience the same degree of deprivation, they would influence the composite deprivation index in exactly the same way as before. Composite deprivation indices which are based upon measures of potential (as opposed to actual) deprivation are consequently potentially misleading if used to monitor changes in deprivation over time; they are also (following a similar logic) problematical if used to compare different areas in a cross-sectional study unless it is assumed that the relationship between potential and actual deprivation is spatially invariant (which it almost certainly is not). The first requirement of a deprivation index is that it should measure the outputs of the process, without confusing the outputs with some of the inputs.

The use of composite indices of deprivation at all must also be questioned. Composite indices are normally justified on the grounds that deprivation is "multidimensional". However, whilst it is accepted that deprivation can take many different forms, it is argued that the consolidation of quite different forms of deprivation into a single (or small number of) composite measure(s) serves only to obfuscate, due to the fact that these different forms of deprivation are the product of different processes which may require quite different policy responses. The problem is compounded by the fact that the weight given to each of the constituent forms of deprivation in a composite index is not determined by any notion of relative importance based upon a theory of deprivation, but is normally determined either arbitrarily (e.g. by assigning each constituent equal weight) or statistically (e.g. using multivariate statistical techniques such as Principal Components Analysis). Principal Components Analysis, for example, assigns weights to each constituent variable based upon its degree of multiple correlation with each of the other variables, but what is the theoretical justification for giving variables which have similar spatial distributions more weight than those with different spatial distributions, especially when the choice of which variables to include in the analysis is often arbitrary and constrained by pragmatic considerations such as data availability?

Multiple deprivation indices do of course measure something vaguely related to deprivation. The problem is knowing what exactly that "something" is. One alternative is to measure different forms of deprivation using different indicators. This would facilitate a clearer interpretation of the causal processes and provide clearer guidelines to the appropriate policy responses. A possible drawback with this approach is that different types of deprivation tend to be found in different types of area, therefore there is a danger that the amount of resources allocated to tackling different forms of deprivation may be influenced by political considerations — i.e. politicians serving predominantly rural areas are likely to favour certain types of deprivation to be tackled with more urgency than those favoured by politicians serving predominantly urban areas. However, at least the process would be more transparent than one based upon an allocation of resources based upon a supposedly objective technocratic identification of areas of need using indices of multiple deprivation which, to a large extent, can be manipulated to produce whatever results the researcher wishes to achieve. The fact that the researcher may not consciously wish to achieve any particular objective does not detract from the essentially arbitrary nature of the whole exercise.

Apart from considerations arising from the choice of variables, consideration must also be given to the limitations imposed by the definition of the areas to which these data refer. In the case of the Small Area Population Statistics, in the absence of information for enumeration districts, the observation units are either DEDs in rural areas or wards in the larger urban areas. There are a number of problems associated with these areas, as noted by Cook *et al.* One is that DEDs vary considerably with regard to population. Areas with very small populations are more likely to be identified as having either extremely low or extremely high percentage rates of deprivation because of what is sometimes referred to as the "small numbers" problem — i.e. one or two additional cases of deprivation can change a small percentage rate into a large percentage rate in areas with a small population (Pringle, 1996). This problem would be serious enough if it was simply a problem of "instability" because every area had a small population, but in the Republic the problem is compounded by the very large variation in size between DEDs. This creates a *systematic* bias in which the small population areas are more likely to identified as suffering the highest levels of deprivation, whereas the more populous areas are much less likely to identified as problem areas, irrespective of the number of cases of deprivation they contain. This is most clearly seen in towns like Dundalk, Portlaoise and Newbridge which fall within a single DED. Even if these towns contained quite extensive areas of multiple deprivation, they would not be picked up by the deprivation indicators because the cases of deprivation would be disguised by being mixed in with a larger non-deprived population.

Cook *et al.* also draw attention to what they refer to as the settlement size bias. Even if all the areas contain a similar sized population (as they do in Northern Ireland), a settlement size bias may arise because the subdivisions of the larger towns and cities are more likely to be socially homogenous, because of socio-spatial segregation, than those in the smaller towns and villages. Areas within the larger towns and cities will therefore tend to score either very high or very low on the deprivation indicators, whereas those in the smaller towns and villages will more likely lie somewhere in between. This in turn will create a bias in favour of areas within the larger towns and cities if deprivation indices are used to identify the areas which have the highest rates of deprivation. Similar effects have been observed in comparisons of educational disadvantage between schools (Walsh, in this volume).

There are many other problems associated with analysing spatial data (such as the ecological fallacy and the modifiable unit problem) with which geographers are very familiar (Robinson, 1950; Openshaw, 1984). There is no need to document these here, but it should maybe be pointed out that one should not assume that these problems are equally familiar to everyone. Likewise the ways in which the findings of a study are presented may also be potentially misleading. For example, geographers know almost instinctively when reading a choropleth map that many of the subdivisions which are largest in areal extent contain very small populations, and that the highest concentrations of people are often found in the areas which are smallest in spatial extent. However, it must not be assumed that non-geographers will read the map in the same way. It is therefore necessary to supplement choropleth maps with a descriptive text to aid interpretation.

THE GEOGRAPHY OF POVERTY IN IRELAND

Taking account of the reservations above, what can we say about the geography of poverty and deprivation in Ireland? The maps of deprivation indicators for small areas (as, for example, in the chapters by Haase, Williams *et al.*, and Cook *et al.*) suggest that the large scale pattern of deprivation is complex and that whilst some general tendencies may be identified, such as the circles of affluence around the major cities noted by Haase, the level of deprivation may also vary markedly between adjoining small areas. It is also clear that different forms of deprivation have different spatial distributions (e.g. unemployment tends to be more prevalent in urban areas, whereas problems associated with age dependency tend to be more prevalent in rural areas). The reader is referred to Chapters 2 to 4 for further details on the large scale distribution of deprivation; this section will confine itself to a few more general observations.

The first point that must be made is that, although deprivation may be concentrated in localised pockets at large scale, the distribution of poverty and deprivation at the national level is much more uniform than one might imagine. For example, Williams *et al.* in an analysis of the 1994 Living in Ireland Survey found the risk of deprivation was almost identical in 6 of the 9 planning regions in the state. The North-West had a higher rate (as one might expect), whilst the West and North-East regions (perhaps more surprisingly) had rates below the national average. However, the most significant feature of these findings is not that the North-West is more deprived than other parts of the state, but that the differences between the other regions are so small. The risk of deprivation, in short, is more or less the same in almost all regions of the state. This is an interesting finding given the recent debate on regionalising the state in order to retain EU Objective One funding — especially as the proposed new "Objective One" region contains the two planning regions which the Living in Ireland Survey suggests have the lowest risks of poverty.

William *et al.* also found that the risk of poverty was spread over all settlement types, although higher risks were found for small towns (i.e. with populations less than 3,000) and other county boroughs (i.e. excluding Dublin). This is an important finding because public discussion about poverty and deprivation often tends to be polarised into an urban versus rural debate based upon stereotypes such as isolated small-farming communities in the West or the "urban jungles" in the Dublin new towns. Both types of area clearly have problems which need to be addressed, but the survey results indicate that there is obviously a lot of people with equally serious problems living in areas which do not conform with either stereotype. Indeed, the findings from the Living in Ireland Survey suggest that the risks of poverty are actually greater in settlements which do not conform with the stereotypes (although such conclusions need to be treated with some caution bearing in mind the modifiable areal unit problem and other types of methodological problem discussed by Cook *et al.*).

Tenure type would appear to be the best single predictor of a high risk of poverty. Williams *et al.* found that the risk of poverty is approximately six times greater amongst those living in houses rented from a local authority than amongst those who own their own house. This is not too surprising when it is borne in mind that social housing is allocated is accordance with need. Williams *et al.* also found that the risk of poverty for those living in local authority housing was remarkably similar in different types of settlement. In fact, the risk of poverty was almost identical for those living in local authority housing whether it was in small towns (i.e. with populations less than 3,000), in the other county boroughs, or in Dublin city and county. It was somewhat lower, although still high, amongst those living in the larger towns (i.e. with populations more than 3,000), but significantly lower amongst those living in local authority houses in open country. Indeed, the risk of poverty for people in local authority housing in open country was no worse than the average for the country as a whole, whilst that for people in private housing in open country was considerably lower than the national average. These findings suggest that the largest concentrations of poverty and deprivation are located in local authority housing estates in towns and cities of all sizes. However, it is also important to note (especially if formulating policy responses) that, as indicated in Table 3.2, approximately 50 per cent of all cases of poverty are located in non-local authority housing.

One final point, made by several authors in this volume, is that cross-sectional surveys and maps of poverty and deprivation do not tell the full story because they only provide a "snapshot" which may give a misleading impression by ignoring the temporal dimension. For example, unemployment measures frequently understate the limitations upon employment opportunities in rural areas because of out-migration — i.e. an area may have a relatively low unemployment rate because most of those who could not find suitable employment have left the area to seek employment elsewhere. Thus, what on the surface appears to be a relatively healthy situation, may in fact disguise a serious problem with regard to employment opportunities, and possibly even with regard to the survival of the whole community. The unemployment rates in urban areas may, by the same token, be inflated by an increase in population due to an in-migration of rural people seeking work — i.e. the situation in urban areas, with regard to employment opportunities, may not be as bad as the unemployment rates suggest. However, this needs to be qualified by a consideration of who actually gets the better jobs. If the better jobs in the large cities are being taken up by better educated people from rural areas (who had to leave home because of the lack of suitable opportunities), then the fact that there are jobs available for qualified people in urban areas may not be of much benefit to urban people who lack the qualifications. To put it more bluntly, who is worse off: those who have to leave home to find suitable employment; or those who will never find decent employment whether they leave home or not? This is not a question we propose to attempt to answer: we simply raise it to highlight the need to be more conscious of the need to consider space as not simply as a blank area to be coloured in on a map, but as the context for lived experiences.

GEOGRAPHY AND THE EXPERIENCE OF DEPRIVATION

The likelihood of being objectively deprived would seem to be related to geographical location; however, if one is objectively deprived, is one's subjective experience of that deprivation influenced by where one lives? Tovey suggests that "rural poverty in Ireland is quite unlikely to be experienced in the same way as rural poverty in England" due to the way in which rural life in England is idyllicised as desirable and good, whereas rural dwellers in Ireland "have seen their way of life systematically devalued and delegitimised". One's perceptions are also likely to be influenced by whether one deliberately opted for a spartan "alternative" rural way of life following a conscious rejection of the materialism of the urban way of life, or whether one was left stranded in a declining rural community by the loss through out-migration of one's friends and relations. Similar levels of "objective" deprivation may therefore be experienced differently depending upon one's personal circumstances. They may also be experienced differently by people living in different locations. Several authors in this volume have touched upon the influence of location upon one's subjective experience of deprivation, but the subject is one which obviously requires further investigation.

It is not difficult to imagine how the experience of deprivation might be influenced by location. For example, the absence of access to private transport will obviously have a bigger impact upon those who live in less accessible locations and to those who do not have access to public transport. Non-ownership of a car will therefore on average be a more serious problem for rural dwellers than for those living in towns. However, the impact will also be influenced by other factors. For example, if one has car-owning friends who live nearby, the effects of not owning a car will not be as serious as they would be for someone who either has few friends or whose friends do not own cars either. The latter situation frequently arises, especially for women with young children, in the new suburban estates on the edges of the larger cities. The fact that one may only be a few miles from shops and other facilities is not much consolation if one does not have access to either private or public transport — it does not make a big difference if you are two miles from the shops or twenty if you have young children but no transport. Accessibility has an important bearing upon the quality of life, but it does not receive as much attention as it should.

The experience of deprivation is also likely to be influenced by the degree of social mix. If one is poor and living in an area which is deprived, the experience of poverty is likely to be quite different to being poor and living in an area which is socially mixed. People living in a deprived area not only have to contend with problems arising from their own personal circumstances, but also with the problems associated with living in a deprived area (such as crime, drugs, social stigmatisation, etc. as described in this volume by Bartley and McCafferty). On the other hand, the subjective experience of deprivation may be more acutely felt by deprived people living in a socially mixed area because they will be more conscious of the lifestyles of their more prosperous neighbours. The age structure of an area may likewise impact upon the quality of life: for example, a deprived area with a large percentage of teenagers will probably be perceived as a more threatening environment by elderly people than a similarly deprived area dominated by "empty nest" households.

The experience of deprivation is also likely to be influenced by less tangible factors, such as the degree of community spirit. "Objective" indicators of deprivation take little account of the non-material factors which contribute to the quality of life, such as whether one has "good neighbours" who will rally round in time of need. The major problem in this regard is that a strong sense of community only develops over a very long period of time, but it can be destroyed in a very short time; and once destroyed, it may never be replaced. The out-migration of young people from rural areas seeking employment opportunities results not only in those people suffering dislocation as they are torn from their friends and relations, but it also has a psychologically debilitating effect upon those who are left behind, resulting in extreme cases in communal demoralisation. It is essential that in our quest to minimise material deprivation (e.g. by adopting policies which maximise aggregate economic growth by focusing upon selected urban centres) we do not overlook the non-material implications which the movement of people from rural areas may have upon the communities they leave behind.

It is also important to recognise that not all vulnerable communities are located in rural areas. The fact that so many people have been forced to leave established rural communities for the anonymity of new residential areas in the larger urban areas has perhaps contributed to a belief that urban areas do not have strong communities. However, the older neighbourhoods in urban areas often have a very strong sense of community, as anyone who has ever lived in these areas will attest. The communities in these neighbourhoods, many of which are working class, are sometimes threatened by redevelopment, especially in inner city areas. Planners are more enlightened than they were in the past, and are now less likely to relocate inner city communities in the suburbs, but many inner city communities now find it difficult to resist the encroachment of commercial and other non-residential land-uses as the central business district expands, or the invasion of new young middle class residential population following gentrification, resulting in tensions between the older residents and the newer ones isolated behind security fences (KPMG, *et al.*, 1996). The destruction of these communities, many of which are inhabited by elderly residual populations characterised by high levels of material deprivation, is every bit as serious a problem as the destruction of rural communities.

GEOGRAPHY AND THE CAUSES OF POVERTY

The fact that the likelihood of poverty and deprivation is significantly greater in some areas compared with other areas suggests that the processes governing the spatial distribution of poverty and deprivation do not operate in a geographically random manner — i.e. the reasons why some areas experience more poverty and deprivation than others can not simply be attributed to "bad luck" or chance. If we can develop our understanding of the nature of the factors and processes which shape the geography of poverty and deprivation, then it may be possible to formulate more effective policies for tackling these problems.

There are many different forms of deprivation, each of which is the product of a different mix of causal factors. Poverty, as indicated in Figure 14.1, is a major factor underlying many forms of material deprivation, but other factors also play a role. Non-material deprivation is the product of an even more diversified set of factors. No attempt is made here to develop a comprehensive theory of the geography of poverty and deprivation: much more research is required before this will be possible. Instead, we will confine our comments to a few more general points.

It is important to make a distinction between two types of process which can affect the geographical distribution of poverty and deprivation. The first set of processes is those which increase the relative risks of individuals living in one area experiencing poverty or deprivation compared with individuals living in other areas — i.e. the risk of poverty or deprivation is a function of geographical location. We shall refer to these as *geographical causal processes*. The second set of processes influence the likelihood of individuals who experience poverty or deprivation being located, or even concentrated, in particular areas, but do not directly influence the likelihood of an individual experiencing poverty or deprivation. We shall refer to these as *geographical distributional processes*. Geographical causal processes influence both the quantity and spatial distribution of poverty and deprivation, whereas distributional processes only influence the spatial distribution of problem cases — they do not have a major influence on the total number of cases.

Geographical Causal Processes

Geographical causal processes influence the likelihood of poverty or deprivation, but vary in intensity over space. For example, land quality and farm size both influence the likelihood of being able to maintain a sustainable income in agriculture, and both tend to vary regionally — i.e. there is a higher concentration of small farms and farms on poor land in the North-West of the country compared with the South-East, therefore (all other things being equal) one would expect higher levels of rural poverty in the North-West. Likewise, given that building standards have tended to improve over time, the likelihood of housing deprivation (e.g. the absence of amenities, such as an indoor toilet) in an area is likely to be at least in part a function of the average age of the housing stock, which varies from area to area.

Geographical causal factors also include distance, especially as measured in terms of travel time and travel costs. Distance is a major determinant of accessibility, which in turn is a major determinant of several forms of deprivation (especially non-material forms of deprivation). Some of the implications of inaccessibility for the provision of services to people in rural areas are graphically described in this volume by Cawley. Poor accessibility tends to disadvantage people in rural areas. However, the problems of inaccessibility are intensifying because service provision is being reduced in some rural areas due to economic rationalisation against a background of population decline. Increased inaccessibility results in a decline in social interaction, which in turn undermines the sense of community and reduces the likelihood of younger people wishing to remain. This in turn contributes to the likelihood of further population decline, reduced interaction and a diminution in the quality of life for those who remain. And so the downward cycle continues.

Geographical Distributional Processes

Geographical distributional processes have the effect of concentrating more needy people into certain areas and more affluent people into other areas. The housing market, for example, exerts a major influence upon the geography of poverty and deprivation, especially within the larger urban areas. Private developers generally prefer to build luxury houses in "exclusive" areas, where the address will add to the prestige and consequently the exchange value of the property; whereas more modest private houses, catering for different sections of the housing market, are located at other locations. Private sector housing, especially that at the upper end of the scale, tends to be located away from public sector housing. Thus, the spatial division of housing type into different areas tends to results in a spatial segregation of occupants by income groups. The most expensive houses will tend to be occupied exclusively by those on a high income, excluding those on a more modest income; the more modest private housing will tend to be occupied by people on a more modest but steady income; whilst people on lower, or less dependable incomes, will tend to be excluded from the private housing sector altogether because of a reluctance by the lending institutions to lend money to people regarded as bad risks. The housing market therefore tends to sort people into different areas according to their financial circumstances; those who are most susceptible to poverty will tend to be sorted into public sector housing areas or the lower end of the private rented sector. Poverty and deprivation will consequently tend to be concentrated into particular areas, not because these areas necessarily contribute to the risks of someone living there becoming deprived, but because people who are either deprived or at risk of deprivation are excluded from other areas. These processes are possibly more pronounced in Ireland because of the high level of home ownership and the ways in which housing finance operates, resulting in a high degree of socio-spatial polarisation.

It should be noted that there may also be a "pecking order" within the local authority housing sector itself. Some estates are perceived as much less desirable places to live than others, either because of the quality of the housing stock or for social reasons (e.g. stigmatisation, anti-social neighbours), consequently many of the residents living there may seek a transfer to a different estate. The net effect of this is that the more desirable housing estates over the course of time develop a settled community, characterised by higher levels of tenant purchase, whilst the less desirable ones will have a high turnover of population as people move out and are replaced by new tenants (as, for example, in Southill, as discussed in this volume by McCafferty). Also, although most local authorities would deny it, there is a strong suspicion that some estates are *de facto* "written off" by the housing authorities and become a "dumping ground" for vulnerable households — e.g. by refusing applications from tenants in these estates regarded as bad risks for a transfer to a different estate. The operation of a policy of this nature, whether formal or informal, will obviously tend to result in a spatial concentration of cases of poverty and deprivation.

It should also be stressed that the very worst quality housing is usually in the private rented sector rather than the public sector. Although the quality of public sector housing varies from estate to estate (and even within estates), public sector housing is generally built to reasonably high and consistent standards. Private rented accommodation, on the other hand, embraces the complete spectrum from luxury penthouse suites to slum housing. The latter provides accommodation for students and other young transient populations, but it also acts as a more permanent home for those who fail to qualify for local authority housing or, increasingly, those regarded by the housing authorities as problem cases. The very worst housing conditions, although typically small in spatial extent, are frequently found in private sector housing. However, it is important to stress that, although slum housing is a form of deprivation, it is not the cause of the poverty of the people who inhabit the housing; they would be still be poor even if they were moved to better accommodation.

Scale and Context

Different geographical causal and distributional processes operate at different spatial scales — i.e. factors which are important to the understanding of the spatial distribution of poverty and deprivation at one spatial scale are not necessarily important at other spatial scales. European initiatives and government policies, for example, play a major role is determining the distribution of wealth in Ireland at a broad regional level, but local authority planning departments and housing departments may be much more influential in determining the geography of poverty at local level (e.g. by permitting or rejecting housing developments in particular areas; or by policies adopted for the management of estates). However, the fact that the impact of different factors tends to be more noticeable at different spatial scales does not mean that outcomes at one spatial scale are unrelated to processes operating at quite different scales. For example, the closure of an established factory in a medium sized town can have a devastating impact upon the local economy and the morale (and even the health) of the local population; however, the closure of the plant may have nothing to do with factors originating at the local level. The plant, if part of a transnational corporation, may have been closed because of circumstances dictated at the corporate headquarters in the US or Japan. Even if locally owned, closure of the plant may have been dictated by the loss of overseas or home markets by increased competition from abroad. Likewise, the government's ability to tackle problems at a local level may be constrained or driven by European policy. We live in an increasingly globalised world, in which decisions taken at one spatial scale may have ramifications at other spatial scales. It is very important, when attempting to untangle the causal and distributional influences upon the geography of poverty and deprivation, to correctly identify the scale and origins of the problem: there is little point in attempting to solve a problem manifest in a local area, if the sources of the problem originate elsewhere or at a different spatial scale.

Geographers are perhaps guilty of focusing almost exclusively upon processes which operate at the global, national and inter-regional scales, in the quest for a "big theory", whilst to some extent neglecting processes at the local level — in particular, those which contribute to a sense of place. Each place is unique: it occupies a unique position relative to other places, it has a unique history, and the people living there have a unique local culture (i.e. a unique set of collective memories and often a similar outlook). It is important when analysing maps showing the distribution of poverty or deprivation to see the maps as more than a pattern of relative highs and lows, but as assemblages of places which provide the context for the lived experience of different groups of people. Several essays in this volume suggest that place-related factors may have a major influence upon the ways in which deprivation is experienced, whilst some (most notably those by Bartley and McCafferty) argue that place-related factors may also influence the extent of objective deprivation in an area.

The way in which processes operate may also vary depending upon the context. Factors which may be important in an urban context are not necessarily important in a rural context, and *vice versa*. For example, housing tenure is generally a fairly reliable indicator of potential deprivation in urban areas, but much less so in rural areas; car ownership is an important indicator of the likelihood of accessibility-related forms of deprivation in rural areas, but less so in urban areas. The point is not that certain causal factors are more common or less common in urban and rural areas; it is that the importance of these factors to an understanding of poverty and deprivation is different in different spatial contexts. The example given contrasts urban and rural areas, but similar variations may occur between regions or between different types of sub-areas within major cities. If, as suggested, the laws governing poverty and deprivation vary between different types of area, then this reinforces the need for deprivation indicators to measure only the outcomes of the processes (i.e. actual deprivation, whilst excluding potential deprivation). The inclusion of measures of potential deprivation (e.g. percentage of small farms, car ownership) will only have relevance in areas where that particular factor has a strong influence upon the likelihood of actual deprivation (e.g. rural areas).

The fact that the processes leading to poverty and deprivation in different spatial contexts may be different should not be interpreted as indicating that they are independent of one another. In much the same way that processes operating at one spatial scale can affect outcomes at a different spatial scale, processes operating in one spatial context may be influenced by processes originating in quite different spatial contexts (and also, quite often, different spatial scales). Tovey in this volume, for example, explains how the production of poverty in rural areas is conditioned by external forces, such as changes in the global food system, industrial urbanisation, and new forms of rural resources use (such as tourism) prompted by external demand.

Structuralist Interpretations
Several authors in this volume, most notably MacLaran and Tovey, have stressed the importance of structural processes. Poverty is a function of social inequalities, and social inequalities are an essential feature of the capitalist mode of production — i.e. it is argued that poverty (defined in relative terms) is not simply an unfortunate by-product caused by a malfunction in the system; rather, it is an unavoidable outcome of the way in which the system operates.

Most people in Ireland have sufficient resources to provide for the basic necessities of life (i.e. food, shelter, clothing), therefore poverty is normally defined in relative terms — i.e. people or households are defined as being in poverty if their income is less than a given percentage of the national average. Relative poverty is unrelated to the absolute amount of resources available to poor people, but rather is defined solely in terms of how the resources available to poor people compare with those available to others. The percentage of people in relative poverty consequently is not a function of the national wealth, but rather is a function of how that wealth is *distributed* throughout society. It is quite possible for a wealthy society to have a high percentage of people in relative poverty (e.g. the United States), and for a less affluent society to have a much lower percentage of people in poverty. The processes which determine the extent of poverty in a society are those which determine how wealth is distributed rather than those which govern how wealth is created — i.e. the key processes governing the number of people below the poverty line in a society are more a function of social structures than economic factors.

Relative poverty is intimately related to social inequality. Structuralists argue that it is impossible to eliminate social inequalities in advanced capitalist societies because inequalities are essential to the operation of the capitalist mode of production. Put crudely, value is created under capitalism through the expenditure of labour power; by paying workers less than the value of the goods they produce, the owners of capital are able to make a profit on their investments. Capital accumulation is therefore based upon an exploitative relationship, which in turn is dependent upon social inequalities. The unemployed play an indirect role in maximising profits by helping to keep wage levels low. In the absence of social inequalities there would be no potential for one class to exploit others, investment would cease and the whole system would grind to a halt. It is therefore impossible to eliminate social inequalities without fundamentally transforming the essentially exploitative nature of the capitalist mode of production.

Structuralists argue that the sources of poverty and social inequality are ultimately rooted in the structures of capitalist society, but this may be disguised by the fact that poverty appears to be associated with particular local areas. However, localised pockets of poverty and deprivation are not the product of processes operating solely within these localised areas; rather they are the outcome of the processes which produce and reproduce social inequalities within broader society — i.e. they are the local manifestations of processes operating at a system-wide or global scale. Local areas provide the context within which the processes of global capitalism become manifest. Different areas provide different contexts, due to differences in natural resources (e.g. land quality, scenic attraction) modified by the legacy of history (e.g. colonialism and landholding, past economic development, past migration histories, existing infrastructure). Place therefore acts as a mediator of global processes, resulting in the complex mosaic depicted on maps of poverty and deprivation.

POLICY IMPLICATIONS

Structuralist interpretations could be misinterpreted as an argument for doing nothing. If social inequalities are an essential feature of the capitalist mode of production, then it might appear that there is not very much that anyone can do to bring about the type of fundamental transformation in the system that would be required to eliminate social inequalities. However, although it may be impossible to eliminate social inequalities, there are reasons for believing that it may be possible to influence the *extent* of social inequalities and hence the number of people living in poverty. Some advanced capitalist societies (e.g. Japan and Sweden) are much more egalitarian than others (e.g. US, UK or, for that matter, Ireland) (Judge, Mulligan and Benzeval, 1998; Wilkinson, 1992; Wilkinson, 1996). The degree of income inequality within countries does not appear to be related to either national wealth (as measured by GNP per capita) or to geopolitical influence, but it would appear to reflect the prevailing political culture within countries with regard to social issues. Given that political outlooks can be modified by persuasion, there are grounds for believing that it may be possible to influence the extent of income inequalities within societies by promoting and adopting more egalitarian policies at national and possibly even supranational (e.g. European Union) scales. It might also be noted parenthetically that, apart from any ethical or social arguments which might be made in favour of increased equality, or the need to reduce the social costs of inequality, there is some evidence to suggest that the more egalitarian societies have experienced faster rates of economic growth than those with larger income disparities in recent decades (Birdsall, Ross and Sabot, 1995). Policies aimed at increased egalitarianism could therefore benefit more than just the most needy sections of society.

The policies required to ameliorate the structural processes leading to social inequalities need to be initiated at a national or supranational level. It is envisaged that many of these policies would be primarily redistributive, using instruments such as taxation and welfare policies to reduce the gulf between the "haves" and "have nots". Where possible government policy should also aim to foster social inclusion. This entails more than simply providing additional resources for disadvantaged groups (although this is an essential prerequisite) — it entails providing a mechanism whereby the disadvantaged can participate more effectively in all aspects of life, including the decision making process. Although the required changes will need to be initiated at a national level, their successful implementation requires changes to be made at other spatial scales. Decision making in Ireland has traditionally been strongly centralised and "top down" in nature (although, because of the nature of the electoral system in Ireland, the top decision makers are arguably more accessible and answerable than in most European countries). There is therefore a need to overhaul the administrative system at large. Walsh (Chapter 6) argues, in the context of rural development, for a stronger "bottom up" input through partnership schemes, and greater vertical and horizontal co-ordination between agencies at different levels and in different sectors. Many of the points made by Walsh have equal relevance for urban development and could also be applied to the social arena.

Several authors in this book have discussed the pros and cons of local area-based strategies as a means of improving the efficiency and effectiveness of present policies with regard to: a) providing additional resources and support services to those most in need; b) counteracting the processes which concentrate poverty in local areas or which have a negative influence upon the way in which poverty is experienced in particular locations; c) improving the delivery of services through better co-ordination between agencies, better adaptation to local circumstances, and innovation and experimentation; and d) promoting social inclusion by facilitating local input into the decision making process. There are clearly difference of opinion between different authors regarding the usefulness of local area-based strategies. The remainder of this section reviews some of the points made with regard to each of the four objectives identified above.

Targeting Additional Resources

Given that it is possible to identify localised pockets of poverty and multiple deprivation, one obvious argument in favour of local area-based schemes is that they provide a convenient mechanism for channelling additional resources and services to people in greatest need. However, several authors in this volume have identified problems with this strategy. One set of reservations relate to the way in which the areas designated to receive additional resources are identified. Whilst most of the areas which have been designated for area-based schemes can be justified on the grounds that they experience deprivation, in most cases there does not appear to have been a systematic attempt to ensure that the areas designated are the *most* deprived areas. Indeed, it is debatable as to whether it would even be possible to identify the areas of greatest need given the restricted nature of the type of information available on poverty and actual deprivation (in particular the absence of any information available on income) from the census (the only comprehensive source of data available for small areas). The comparatively large size of the "small areas" for which these data are available (especially for urban areas outside the five county boroughs) and the massive disparities in the size of these areas (in terms of population) raise other fundamental doubts about the validity of any exercise to identify the areas of greatest need, as noted by Cook *et al*.

A second set of reservations relate to the efficiency of using area-based schemes as a mechanism for delivering resources to people in need. As several authors in this volume have noted, severely deprived people may represent only a minority of those living in deprived areas, therefore a substantial number of people who do not require assistance could benefit (depending on the uses to which the resources are put) from living in a designated area. More seriously, deprived areas generally contain only a small percentage of the total number of deprived people in the state, therefore the provision of additional resources to people in designated areas will only benefit a small percentage of those whom one would ideally like to assist. Indeed, even if one was to designate every local authority housing estate in the country for special attention (i.e. 18 per cent of all households in the state), the incidence figures reported by Williams *et al*. suggest that the designated areas would only include about 50 per cent of all those in poverty.

The above reservations do not provide overwhelming arguments against the use of positive territorial discrimination, but they do suggest that if a strategy of positive territorial discrimination is to be adopted then more thought needs to be given to the objectives in identifying areas for special attention and related issues, such as what size these areas should be, what level of resources need to be provided, and how many areas need to be designated to have an optimal impact (see Walsh, chapter 13). More attention should also be given to the methods used to identify the areas to be designated.

Counteracting Local Processes
Given that the risks of poverty and deprivation appear to be especially high in particular local areas, a second possible application

of a local area-based strategy is to identify and counteract the local processes which contribute to this increased risk. It is probably fair to say that the authors in this volume are divided in their opinions on the usefulness of this particular strategy. On the one hand, it is not difficult to identify factors which operate at a local level and which contribute to the likelihood of individuals living in the area experiencing poverty and deprivation (such as poor primary and secondary level schools, poor access to employment opportunities, the stigma attached to specific areas, etc.). It is also easy to identify local factors which influence the experience of deprivation in particular areas, resulting in a reduced quality of life (such as poor housing quality, poor public transport services, absence of shops and services, and a depressing "run-down" environment, etc.). The effects of such factors could be ameliorated by local development strategies. On the other hand, others (especially those informed by a structuralist analysis) would argue that, whilst the provision of additional resources to counteract local processes could make a substantial difference to the quality of life in the designated areas, tackling the local "causes" of deprivation is ultimately futile because it ignores the real causes of social inequality which operate within broader society.

It is also important when tackling the causes of local concentrations of poverty to make a distinction between factors which increase the risks of poverty and deprivation for individuals living in particular areas (geographical causes of poverty) from those which increase the likelihood of the poor and deprived living in particular areas (geographical distributional processes). For example, the renewal of a deprived inner city area could result in the elimination of a blighted black-spot, but if the former residents are simply displaced to a different location (or, even worse, scattered between different locations) to make way for commercial land-uses or a new set of more affluent residents, then the elimination of poverty in this particular area will have done very little to counteract poverty and deprivation overall — it will simply result in the poor being relocated (quite possibly to an area where, because of dislocation from friends and familiar surroundings, their quality of life may be even lower than it was previously). This highlights the importance of placing local interventions in a wider policy context, such as social housing policy, reform of the Common Agricultural Policy, and urban regeneration policy.

If pursuing a local area-based strategy to counteract local causal processes, then a great deal of thought needs be given to the level of resources which would be required. Few areas have sufficient financial mechanisms to retain local investment, with the result that such resources as there are (e.g. social welfare transfers) simply drain out of an area. The strengthening of the local financial infrastructure (e.g. community credit unions) needs to be given greater emphasis in area-based schemes. The major problem with area-based schemes in the past is that the level of funding has been totally inadequate compared with what would be required to make a real and meaningful impact within the designated area. Under such circumstances, area-based schemes have made only a marginal impact and could arguably be claimed to be counter-productive because they help reinforce the illusion that the problems found at local level have local causes that are amenable to local intervention, thereby distracting attention away from the more fundamental causes of social inequality and deprivation in society as a whole. The problems of deprivation consequently tend to be sidelined as problems of particular locations which require special attention, rather than as problems requiring major structural reforms at a national or supranational level.

Improving Service Delivery
Given the highly centralised nature of the Irish political and administrative system, several authors in this volume have suggested that the delivery of services could be improved using local area-based schemes. There are several arguments in this regard. One is that the services which impact upon the quality of life for the more deprived sections of the community originate from within several different government departments, each of which has its own set of priorities and agenda. The net effect is that the delivery of these services is not always as effective as it might be: similar types of service may be duplicated by different departments, whilst other types of service may be neglected because it is not clear within whose area of responsibility they should fall. It is argued that local area-based schemes provide a mechanism for

greater local co-ordination of services originating from different sources.

A second line of argument is that local area-based schemes provide a more flexible framework for service delivery — i.e. the management of each local-area scheme could tailor the provision of services to specific local conditions. For example, small farmers in the South-East and North-West regions experience problems which could be alleviated by interventions from central government. However, given the differences in landholding structures, land quality, and the types of agriculture practised, the problems experienced by small farmers in the two regions are quite different. If developing a scheme to alleviate the problems of small farmers, it would therefore be beneficial to have the flexibility to develop different strategies in different parts of the country. Similar arguments apply with regard to addressing problems of deprivation in urban areas experiencing very rapid urban growth, but poor service provision, and urban areas experiencing stagnancy and decline because of the collapse of the traditional local economy. The required flexibility could of course be accomplished by a decentralisation of decision-making within the relevant government departments, but a local-area based strategy facilitates the possibility of greater co-ordination between departments.

A third set of arguments, arising out of the flexibility provided by adopting an area-based strategy, is that area-based schemes have the potential to be more innovative and might even have a experimental component incorporated in their terms of reference.

One important issue which needs to be resolved, especially given the low level of power devolved to local government, is the most appropriate relationship between special local area-based schemes and democratically elected local government bodies. There is a danger that local area-based schemes constituted by central government, if widely adopted, could further undermine the relevance of local government and therefore, in a paradoxical manner, contribute to a further consolidation of power in central government.

Promoting Community Empowerment
The final set of arguments put forward in favour of local area-based strategies is that they provide the potential for increased

local participation in the decision making process (Walsh et al., 1998). This new community development agenda has taken off at a very fast pace, aided and abetted by the EU through the structural funds. Among community advocates, community development projects and local partnership are seen as an opportunity to influence policy-making in an unprecedented fashion to the benefit of the socially excluded. At the same time, central government has supported these new structures as a way of organising public policy in areas of disadvantage and are increasingly using them as a clearing house for a range of new policy initiatives, e.g. with regard to micro-enterprise, child-care, drug rehabilitation, early school leaving, peace and reconciliation, and urban regeneration. However, whilst local area-based strategies have the potential for promoting community empowerment, there is nothing inherent in the concept of area-based strategies to guarantee that they are in any way more socially inclusive than the alternatives. Local area-based strategies may fall anywhere along a spectrum ranging from being totally "top-down" (e.g. where their main function is to co-ordinate the policies of central government departments at local level), to consultative (i.e. providing an opportunity for the community to express their views, without any obligation to take account of these views), to structures providing for community input into policy formulation, to proactive community mobilisation and participation (i.e. "bottom up"). Spatial scale is obviously an important consideration in this regard: the smaller the area, the more potential there is for genuine community-based participation, or at least influence, in the decision process. However, if the area is large (e.g. the size of a county) then pragmatic organisational considerations will require that the views of the community are at best articulated by spokespersons (e.g. elected politicians, local government officials, or informal "community leaders") whose views may or may not accurately reflect the wishes of the community. Even if the area is small, one should not underestimate the effort that may be required to mobilise active participation amongst a deprived population alienated from the mainstream of public life. Social inclusion requires more than simply a right to participate; it requires a much more proactive intervention to try to encourage this right to be utilised.

Community empowerment also depends on the willingness of other agencies (voluntary and statutory) to share power and decision-making with local people. While many agencies are willing to embrace the rhetoric of "community" (e.g. "community schools", "community policing", "community employment"), there is far less evidence that they are prepared to change their practices. This reflects a paternalistic culture in welfare policy, in particular the dominance of professional elites in determining what the poor want. This, of course, includes politicians who, through the practice of clientelism, exert a controlling influence over people's lives. Community development remains outside "mainstream" state and society. It is only recently that the key concepts of active citizenship and civil society have begun to enter public debate, as in the government green paper supporting community initiatives:

> There is a need to create a more participatory democracy whose active citizenship is fostered. In such a society, the ability of the voluntary and community sector to provide channels for the active involvement and participation of citizens is fundamental. An active voluntary and community sector contributes to a democratic, pluralist society, provides opportunities for the development of decentralised institutional structures and fosters a climate in which innovative solutions to complex social problems and enhancements of quality of life can be pursued and realised. (Department of Social, Community and Family Affairs, 1997, p. 24).

Whether this radical agenda will be implemented remains to be seen. The current reforms in local government, such as the establishment of strategic policy committees, present a test case for the commitment of the state to genuine local democracy.

Several authors in this volume have highlighted the concept of "partnership" as an instrument of policy making and delivery. However, partnerships, whilst preferable to an undiluted "top down" structure, do not necessarily provide the degree of local input which one might assume. The concept of "partnership" assumes there is a compromise between different sectional interests (e.g. government, industry, labour, agriculture) and also, in this context, between the "top" (i.e. central government) and the "bottom" (i.e. the community). The existence of a partnership consequently creates a third tier between central government and the community, which in the context of a local area-based scheme may be represented by a group of professional organisers and community representatives who, whilst acting in what they believe to be the best interests of the community, may not necessarily accurately reflect the views of the community (who may in some cases not even know what is going on and may therefore feel more excluded than ever).

The resource demands associated with local partnerships also have a major impact on community organisations, diverting their activities and key personnel into servicing what are in practice large local bureaucracies. This is also an issue for other partners, especially business and trade unions. But at least here the level of expectations as to the potential of local partnerships is far less.

Three fundamental challenges arise in regard to local partnerships as models of local governance in areas of disadvantage. The first is the almost total dependence on external exchequer and EU funding to support their activities. The current level of EU funding will clearly not continue past 2000, no matter what method of regionalisation the government manages to put in place. The partnerships will thus be obliged either to look elsewhere for resources, or else to substantially scale down their activities.

A second difficulty is the limited degree of autonomy afforded to partner state agencies. This is crucial in that it minimises the scope for altering policy at the local level to better reflect the need for social inclusion and social equity. Mainstreaming the policy impact of local experimentation is thus largely dependent on the support of central government, which is by nature much less responsive to local initiatives.

Finally, there is the detachment of local partnerships from local government. Initially this was pursued because of the perceived stagnation and irrelevance of local government to local socio-economic policy. Now, local government is staging a fight-back as it seeks to reclaim its unique mandate in local democracy. There is also a growing awareness of the importance of physical planning in the local development process. How local partnerships are to fare in this new environment of a re-invigorated local government will have a crucial bearing on their future operation.

The concept of a partnership is potentially progressive (although arguably not as progressive as a genuinely local — i.e. smaller areas — and genuinely democratic system of local government), but there is no guarantee that it will be progressive in practice. Indeed, some would argue that partnerships at local level are little more than a device for conflict management and social control. Their main purpose is not to solve problems at local level (where, it is argued, the scope for action is highly constrained), but to defuse potential opposition and to distract attention away from the real decisions taken at national level to promote capital accumulation, at the expense of social inequalities. It is at the national, rather than the local level, where change must be implemented if we are to reduce the social inequalities which give rise to poverty and social deprivation.

Summary

Some of the main conclusions arising from the above discussion are summarised in point form:

- More research is required on the geography of poverty and of different forms of actual deprivation in Ireland. To fully understand the geography of poverty and deprivation, information is required on smaller areas than DEDs, especially in towns where the entire town falls within a single DED.

- In order to explore the geography of poverty (as opposed to deprivation), more detailed information is required on the distribution of income by small areas. This, in theory, could be achieved by the introduction of an income question in the census of population, or by further processing of administrative records. However, it must be conceded that both strategies could be politically controversial.

- When exploring the geography of deprivation, a distinction should be made between potential and actual deprivation.

- If calculating indices of multiple deprivation, the weights applied to the constituent indicators should be based upon a theory of deprivation rather than assigned arbitrarily or determined by the degree of multiple correlation with other variables.

- When attempting to understand the processes influencing poverty and deprivation, it is important to distinguish between the processes which cause poverty and the processes which simply affect its spatial distribution.

- More research is required on the experience of deprivation within different spatial contexts.

- Local area-based responses at the local/community level have much to offer, but only if utilised in a manner which enhances community empowerment.

- If local area-based responses are to be used as a mechanism for targeting additional resources to those in need, equity considerations require that the areas to benefit should be designated using a more objective assessment of real need.

- The problem of poverty can not be solved by local initiatives. The main emphasis should be placed at a national/societal level on reducing the social inequalities which give rise to poverty.

REFERENCES

Birdsall, N., Ross, D. and Sabot, R. (1995) "Inequality and growth reconsidered — lessons from East-Asia", *World Bank Economic Review*, 9(3), 477-508.

Department of Social, Community and Family Affairs (1997) *Supporting Voluntary Activity*, Stationery Office, Dublin.

Judge, K, Mulligan, J. and Benzeval, M. (1998) "Income inequality and population health", *Social Science and Medicine*, 46(4), 567-579.

Kawachi, I. and Kennedy, B.P. (1997) "The relationship of income inequality to mortality: does the choice of indicator matter?" *Social Science and Medicine*, 45(7), 1121-1127.

KPMG, Murray and O'Laoire, NIERC (1996) *Study on urban renewal schemes*, Stationery Office, Dublin.

McLoone, P. and Boddy, F.A. (1994) "Deprivation and mortality in Scotland: 1981 and 1991", *British Medical Journal*, 309, 1465-1470.

Madanipour, A., Cars, G. and Allen, J. (1998) *Social Exclusion in European Cities. Processes, Experiences and Responses*, Jessica Kingsley, London.

National Anti-Poverty Strategy (1997) *Sharing In Progress*, Stationery Office, Dublin.

Nolan, B. And Whelan, C.T. (1996) *Resources, Deprivation and Poverty*, Clarendon Press, Oxford.

Nolan, B., Whelan, C.T. and Williams, J. (1998) *Where Are Poor Households? The Spatial Distribution of Poverty and Deprivation in Ireland*, Oak Tree Press/Combat Poverty Agency, Dublin.

Openshaw, S. (1984) *The Modifiable Areal Unit Problem*, Catmog 38, Geo Books, Norwich.

Pringle, D.G. (1996) "Mapping disease risk estimates based on small numbers: an assessment of Empirical Bayes techniques", *Economic and Social Review*, 27(4), 341-363.

Putnam, R.D. (1995) "Turning in, turning out: the strange disappearance of social capital in America", *Political Science and Politics*, December, 664-683.

Robinson, A.H. (1950) "Ecological correlation and the behaviour of individuals", *American Sociological Review*, 15, 351-357.

Rokkan, S. (1970) *Citizens, Elections, Parties*, McKay, New York.

Room, G. ed. (1995) *Beyond The Threshold. The Measurement And Analysis Of Social Exclusion*, Policy Press, Bristol.

SAHRU (Small Area Health Research Unit) (1997) *A National Deprivation Index For Health And Health Services Research*, Dept. of Community Health and General Practice, Trinity College, Dublin.

Taylor, P.J. and Johnson, R.J. (1979) *The Geography of Elections*, Penguin, Harmondsworth.

Taylor, P.J. (1993) *Political Geography. World-Economy, Nation-State and Locality*, Longman, Harlow.

Walsh, J., Craig, S. and McCafferty, D. (1998) *Local Partnership for Social Inclusion*, Oak Tree Press and the Combat Poverty Agency, Dublin.

Wilkinson, R.G. (1992) "Income distribution and life expectancy", *British Medical Journal*, 304, 165-168.

Wilkinson, R.G. (1996) *Unhealthy Societies. The Afflictions of Inequality*, Routledge, London.